A TEXT BOOK OF

PHYSICS OF NANOMATERIALS

For

T. Y. B. Sc. (SEM. III)

ACCORDING TO NEW REVISED SYLLABUS OF SAVITRIBAI PHULE PUNE UNIVERSITY FROM JUNE-2015

Dr. R. B. BHISE (M.Sc., M.Phil., Ph.D., DCM, DIT)
Head, Department of Physics
B. J. A. C. S. College, Ale
Tal : Junnar, Dist : Pune

Dr. A. B. BHISE (M.Sc., M.Phil., Ph.D.)
Ex-Head, Department of Physics
C. T. Bora College, Shirur
Tal : Shirur, Dist : Pune.

Dr. V. D. Kulkarni (M.Sc., Ph.D.)
Head, Department of Physics
H. R. College, Rajgurunagar
Tal. : Khed, Dist. : Pune.

Dr. H. R. Kulkarni (M.Sc., Ph.D.)
Principal, Jayawantrao Sawant College
of Commerce & Science
Hadapsar, Pune.

Dr. A. P. Zambare (M.Sc., M.Phil, Ph.D.)
Head, Department of Physics
Agasti College, Akole,
Tal. : Akole, Dist. : Ahmednagar

NIRALI PRAKASHAN
ADVANCEMENT OF KNOWLEDGE

N1862

T.Y.B.Sc. PHYSICS OF NANOMATERIALS **ISBN 978-93-89406-80-1**

First Edition : August 2019

Published By :

NIRALI PRAKASHAN

Abhyudaya Pragati, 1312, Shivaji Nagar,

Off J.M. Road, PUNE – 411005

Tel - (020) 25512336/37/39, Fax - (020) 25511379

Email : niralipune@pragationline.com

Printed By :

➢ DISTRIBUTION CENTRES

PUNE

Nirali Prakashan (For orders within Pune) : 119, Budhwar Peth, Jogeshwari Mandir Lane, Pune 411002, Maharashtra
Tel : (020) 2445 2044, Mobile : 9657703145
Email : niralilocal@pragationline.com

Nirali Prakashan (For orders outside Pune) : S. No. 28/27, Dhayari, Near Asian College Pune 411041
Tel : (020) 24690204 Fax : (020) 24690316; Mobile : 9657703143
Email : bookorder@pragationline.com

MUMBAI

Nirali Prakashan : 385, S.V.P. Road, Rasdhara Co-op. Hsg. Society Ltd.,
Girgaum, Mumbai 400004, Maharashtra; Mobile : 9320129587
Tel : (022) 2385 6339 / 2386 9976, Fax : (022) 2386 9976
Email : niralimumbai@pragationline.com

➢ DISTRIBUTION BRANCHES

JALGAON

Nirali Prakashan : 34, V. V. Golani Market, Navi Peth, Jalgaon 425001, Maharashtra,
Tel : (0257) 222 0395, Mob : 94234 91860;
Email : niralijalgaon@pragationline.com

KOLHAPUR

Nirali Prakashan : New Mahadvar Road, Kedar Plaza, 1st Floor Opp. IDBI Bank, Kolhapur 416 012
Maharashtra. Mob : 9850046155; Email : niralikolhapur@pragationline.com

NAGPUR

Nirali Prakashan : Above Maratha Mandir, Shop No. 3, First Floor,
Rani Jhanshi Square, Sitabuldi, Nagpur 440012, Maharashtra
Tel : (0712) 254 7129; Email : niralinagpur@pragationline.com

DELHI

Nirali Prakashan : 4593/15, Basement, Agarwal Lane, Ansari Road, Daryaganj
Near Times of India Building, New Delhi 110002 Mob : 08505972553
Email : niralidelhi@pragationline.com

BENGALURU

Nirali Prakashan : Maitri Ground Floor, Jaya Apartments, No. 99, 6th Cross, 6th Main,
Malleswaram, Bengaluru 560003, Karnataka; Mob : 9449043034
Email: niralibangalore@pragationline.com

Other Branches : Hyderabad, Chennai

niralipune@pragationline.com | www.pragationline.com

Also find us on f www.facebook.com/niralibooks

Dedicated to ...

My Mother Gangubai

Preface ...

We are happy to place this book "Physics of Nanomaterials" in the hands of B.Sc. /M.Sc. students and teachers of Savitribai Phule Pune University and Bharati Vidyapeeth. This book envisages the revised syllabus. The main objectives of book are to introduce the basic physics behind size, effect of nanomaterials and to understand the working principle of equipments used in nanostructures. Students and teachers will gain knowledge of nanomaterials, their properties, their applications, and various growth techniques. Authors have also discussed tools like UV, XRD, SEM and TEM to characterize the nanomaterials. The general impression of physics students is that nanophysics is very difficult. Hence the theory of nanomaterials is prescribed in a simple and lucid manner with the help of neat and clear diagrams. This book gives a brief description of physics of nanomaterials which are chosen as per the latest curriculum. The book includes introduction, methods of synthesis, characterization techniques, properties and applications of nanomaterials. Each chapter is discussed with objectives, diagrams and description. At the end of each chapter necessary short- and long-answer questions are included. We think that the students will find this book most useful.

We are thankful to Mr. Ashok Bodake and his group members for their efforts and encouragement. We are also thankful to Dnyaneshwar Gramonnati Mandal, B. J. College, Ale (PUNE) for their co-operation and encouragement. We also gratefully acknowledge the patience and understanding shown by our family members during this endeavor.

Our sincere thanks to Shri. Dineshbhai Furia, Shri. Jignesh Furia and whole staff of Nirali Prakashan for their co-operation, nice printing, constant encouragement and efforts in bringing out this book.

Any suggestion for improvements of the book shall be most welcome.

Dr. Bhise R. B.

Syllabus ...

T.Y. B Sc. (PHYSICS)

PH-346 (H): PHYSICS OF NANOMATERIALS

1. Introduction to Nanomaterials: **(10 L)**

(a) Introduction to nanosize materials and its structures

(b) Brief history of nanomaterials

(c) Challenges in nanotechnology

(d) Significance of nanosizematerials and its properties

(e) Classification of nanostructured materials

2. Methods of Synthesis of Nanomaterials: **(12 L)**

(a) Bottom-up and Top-down approaches

(b) Physical methods: High energy ball milling, Physical vapour deposition, Ionizedcluster beam deposition, Sputter deposition, Ultrasonic spray pyrolysis etc.

(c) Chemical methods: Colloidal method, Co-precipitation and sol-gel method

(d) Hybrid method: Electrochemical and chemical vapour deposition.

3. Characterization Techniques: **(11 L)**

(a) UV- Visible spectroscopy

(b) X-ray diffraction

(c) Scanning electron microscopy

(d) Transmission electron microscopy

4. Properties of Nanomaterials: **(05 L)**

(a) Mechanical property,

(b) Electricalproperty,

(c) Magnetic property,

(d) Thermalproperty,

(e) Optical property,

(f) Solubility and Melting point.

5. Special Nanomaterials: **(06 L)**

(a) Special Nanomaterials,

(b) Carbon nanotubes,

(c) Quantum dots,

(d) NanocrystallineZnO and TiO2.

6. Applications: **(04 L)**

Nano electronics, Medical, Biological, Automobiles, Space, Defense, Sports, Cosmetics and Clothing industry.

☆☆☆

Contents ...

1. Introduction to Nanomaterials **1.1 – 1.8**

1.1 Introduction
1.2 Challenges in Nanotechnology
1.3 Brief History of Nanomaterials
1.4 Nanosize materials and Their structures
1.5 Significance of nanosize materials and its properties
1.6 Classification of nanostructured materials

2. Methods of Synthesis of Nanomaterials **2.1 – 2.18**

2.1 Introduction
2.2 Methods of synthesis of nanomaterials
2.3 Classification of synthesis Methods of nanomaterials

3. Characterization Techniques **3.1 – 3.8**

3.1 Introduction
3.2 Optical study
3.3 Structural study
3.4 Morphological study

4. Properties of Nanomaterials **4.1 – 4.8**

4.1 Introduction
4.2 Mechanical Properties of nanomaterials
4.3 Electrical Properties of nanomaterials
4.4 Magnetic properties of nanomaterials
4.5 Thermal properties of nanomaterials
4.6 Optical properties of nanomaterials
4.7 Solubility and Melting point of Nanomaterials

5. Special Nanomaterials **5.1 – 5.8**

5.1 Introduction
5.2 Special Nanomaterials
5.3 Carbon Nanotubes
5.4 Quantum dots
5.5 Nanocrystalline ZnO and TiO_2

6. Applications of Nanotechnology **6.1 – 6.15**

6.1 Introduction
6.2 Applications of Nanomaterials
6.3 Applications of Nanotechnology

Previous Years University Question Papers **P.1 – P.3**

☆☆☆

Chapter 1 ...

INTRODUCTION TO NANOMATERIALS

- Introduction
- Challenges in Nanotechnology
- Brief History of Nanomaterials
- Nanosize Materials and Their Structures
- Significance of Nanosize Materials and its Properties
- Classification of Nanostructured Materials

1.1 INTRODUCTION

Nano-science deals with synthesis, characterization, exploration, and exploitation of nanostructured materials. Therefore, Nano-science is the study of the fundamental principles of molecules and structures with at least one dimension roughly between 1 to 100 nanometers. These structures are known as nanostructures. Nanotechnology (NT) is the application of these nanostructures useful in nanoscale devices. The word 'Nano' means one billionth part of some unit scale that is 1 nm = 1/1,000,000,000 m (1 nm $=10^{-9}$ m). One nanometer is approximately the length equivalent to 10 hydrogen or 5 silicon atoms aligned in a line. Thus nanotechnology is the technology of materials dealing with very small dimension materials usually in the range of 1 to 100 nm. When dimension of materials is reduced below 100 nm, then it's mechanical, optical, magnetic, electrical, and all other properties are changed.

To get the sense of nanoscale,

(a) a human hair measures 50,000 nm across,
(b) a bacterial cell measures few hundred nanometer across,
(c) a commercial microchip measures 130 nm across, etc.

Nanomaterials have taken centre stage of materials research today. An exact definition of nanomaterials is still a point of contention. Therefore, Nanotechnology (NT) is defined as the design, characterization, production, and applications of structures, devices, and systems by controlled manipulation of size and shape at the nanometer scale that produces structures, devices, and systems with at least one superior characteristic.

Presently, Nanotechnology (NT) is complex of fundamental and engineering sciences that integrates a physics, chemistry, and biology of nanostructures with a materials science, electronics, and process technologies focused on a comprehensive research of nanostructures, on a development of atomistic physical-chemical processes, self and automatic assembling of nanomaterials and work-pieces using complex probe microscopes combined with other tools, resulted in a fabrication and manufacturing of Nano-devices, Nano-machines, ultra-low integrated circuits, micro optic-electro-mechanical systems and Nano-biorobots, etc.

The new commercial applications for nanotechnology for the next several years include:

1. Advanced drug–delivering systems.
2. Implantable devices that automatically administer drugs and sense drug levels.
3. Medical diagnostic tools, such as cancer tagger mechanisms and lab-on-a-chip diagnostics for physicians.
4. Cooling wafer to replace compressors in cars, refrigerators, air conditioners and other devices, using no chemicals or moving parts.
5. Sensors for airborne chemicals or other toxins.
6. Solar fuel cells and portable power to provide inexpensive and clean energy, etc.

1.2 CHALLENGES IN NANOTECHNOLOGY

We need to learn more about materials and their properties at the nanoscale. Nanotechnology is an emerging and rapidly growing field whose dynamics and prospects pose many great challenges not only to scientists and engineers but also to society at large. Challenges in nanotechnology include the integration of nanostructures and nanomaterials into macroscopic systems that can interface with people.

Following are the issues that examine the challenges:

(i) Philosophical Issues:

- Drexler challenge
- Metaphysics of nanotechnology
- Truth of nanoscopic images
- Dealing with complexity

(ii) Ethical Issues:

- Social and environmental ethics of nanotechnology
- Managing the risks of nanotechnology

(iii) Societal Issues:

- Public discourses on the future
- Visions and public reactions to nanotechnology

Philosophical issues are focused on philosophy of science, metaphysics, epistemology, and complexity. Drexler challenge is comparison between chemical approaches and molecular manufacturing with nanotechnology. Eric K. Drexler pointed out that there is a fundamental difference between both approaches based on the conceptual, methodological, and theoretical grounds. A metaphysical issue of nanotechnology focuses on particularly its relation to nature. The complexity of nanotechnology both regarding its interaction with society and the various research fields involved. The truth of nanoscopic image refers imaging techniques that allowed nanotechnology to emerge. Several images of nanoscale structures have even become public icons of the field. SEM should not be called images because they are not exact representations of reality.

Ethical issues are related to societal issues, but sociological analysis is not the same as an ethical analysis. Social and environmental ethics of nanotechnology provide a comprehensive analysis of what may count as ethical issues of nanotechnology through the complementary perspectives of social ethics and environmental ethics such as the creation of new materials, uncontrollable replicators, human enhancement, and material abundance. Managing the risks of nanotechnology carefully analyzes, assesses and manages its potential opportunities and risks.

Societal issues of nanotechnology are all tied to futuristic and visionary stories about nanotechnology. These have generated hype as well as public hopes and fears. This helps to how the public might react to such nanotechnology visions.

The small size and complexity of nanoscale structures make the development of new measurement technologies more challenging. Nanotechnology challenges are to stimulate research that is imaginative, innovative, timely and all relevant to the safety of nanotechnology. They include:

1. Instruments to assess environmental exposure to nanomaterials.
2. Methods to evaluate the toxicity of nanomaterials.
3. Models for predicting the potential impact of new, engineered nanomaterials.
4. Ways of evaluating the impact of nanomaterials across their life cycle.
5. Strategic programs to enable risk-focused research.

1.3 BRIEF HISTORY OF NANOMATERIALS

The history of nanomaterials began after the big bang when nanostructures were formed. Nanoscaled smoke elements were formed during the use of fire.

However, the scientific story of nanomaterials began after this.

In 1857, the first scientific report was presented by Michael Faraday that is synthesis of colloidal gold particles. Nanostructured catalysts have also been investigated for over 70 years.

In 1940, fumed silica nanoparticles were being manufactured and sold in USA and Germany as substitutes for ultrafine carbon black for rubber reinforcements. Nanosize amorphous silica particles are found and used in many applications like every-day consumer products ranging from non-diary coffee creamer to automobile tires, optical fibers and catalyst supports.

In the 1960 to 1970, metallic Nano-powders for magnetic recording tapes were developed.

In 1976, first time Nano-crystals produced by popular inert gas evaporation technique were published by Granqvist and Buhrman. It was found that the Maya blue paint is a nano-structured hybrid material. The origin of its color and resistance to acidic studies of authentic samples from Jaina Island. The material of amorphous silicate substrate containing inclusions of metal (Mg) nanoparticles is made of needle-shaped clay crystals that form a superlattice with a period of 1.4 nm. The beautiful tone of the blue color is obtained only when both these nanoparticles and the super lattice are present.

In 1990, Nano-phase engineering was rapidly expanded in a growing of structural and functional, both inorganic and organic materials. These materials are allowing to manipulate mechanical, catalytic, electric, magnetic, optical and electronic functions. The production of Nano-phase materials is often based upon the creation of separated small clusters that are fused into a bulk-like material or on their embedding into compact liquid or solid matrix materials e.g. nano-phase silicon. These materials differ from normal silicon in physical and electronic properties. Such materials can be useful to macroscopic semiconductor processes to create new devices.

Now-a-days, when an ordinary glass is doped with quantized semiconductor, it becomes a high performance optical medium with potential applications in optical computing.

1.4 NANOSIZE MATERIALS AND THEIR STRUCTURES

Nanoscale materials are a set of elements which have at least one element with dimensions less than 100 nm. At this scale, nanomaterials have unique optical, magnetic, electrical, or other properties. These properties have the potential for great impacts in electronics, medicine, and other fields. Some nanomaterials occur naturally. Nanomaterials are resources designed at nanometer level to take the advantage of their small size and novel properties. Generally, these properties are not seen in their conventional and bulk counterparts. Nanomaterials at the nanoscale can have different properties with increased relative surface area and new quantum effects. They have greater surface area to volume ratio than their conventional forms. It leads to the greater chemical reactivity and affect their strength. A nanoscale quantum effect of materials is important to determine the properties and characteristics, such as optical, electrical and magnetic behavior of materials. The range of nanoscale materials commercial products is very broad such as resistant textiles, cosmetics, sunscreens, electronics, paints and varnishes. Nano-coatings and Nano-composites are finding uses in diverse consumer products like windows, sports equipments, bicycles and automobiles. The novel UV-blocking coating on glass bottles protects drinks from the damage by sunlight. Nanoscale titanium dioxide is used in cosmetics, sun-block creams and self-cleaning windows. Nanoscale silica is used as filler in a range of products, including cosmetics and dental fillings.

1.5 SIGNIFICANCE OF NANOSIZE MATERIALS AND ITS PROPERTIES

People are now interested in the nano-dimensional materials having size less than 100 nm and their properties are different from those at a larger scale. Nano-science is a study of phenomenon and manipulation of materials at atomic, molecular, macromolecular scales and their differing properties. Nanotechnology is the design, characterization, production and application of structures, devices and systems by controlling shape and size at the nanometer scale.

Nanomaterials have the structural features in between those atoms and the bulk materials. While most micro-structural materials have similar properties to the corresponding bulk materials. The properties of materials with nanometer dimensions are significantly different from those of atoms and bulk materials due to

(i) the nanometer size of the materials;
(ii) large fraction of surface atoms;
(iii) high surface energy;
(iv) spatial confinement; and
(v) reduced imperfections

Due to small dimensions, nanomaterials have extremely large surface area to volume ratio which makes a larger surface or interfacial atoms. As a result more surface dependent material properties are seen. If the sizes of nanomaterials are comparable to length, then the entire material will be affected by the surface properties of nanomaterials. This in turn may enhance or modify the properties of the bulk materials. For example, metallic nanoparticles can be used as very active catalysts.

Chemical sensors from nanoparticles and nanowires enhance the sensitivity and sensor selectivity. The nanometer feature sizes of nanomaterials have spatial confinement effect on the materials, which brings the quantum effects.

The energy band structure and charge carrier density in the materials can be modified by their bulk and it modifies the electronic and optical properties of the materials. For example, lasers and light emitting diodes (LED) from both of the quantum dots and quantum wires are very promising in the future optoelectronics.

High density information storage using quantum dot devices is also a fast developing area.

Reduced imperfections are also an important factor in determination of the properties of the nanomaterials. Nanostructures and Nanomaterials favor in self-purification process in that the impurities and intrinsic material defects will move to near the surface upon thermal annealing. This increased materials perfection affects the properties of nanomaterials. For example, the chemical stability for certain nanomaterials may be enhanced. The mechanical properties of nanomaterials will be better than the bulk materials.

The superior mechanical properties of carbon nanotubes are well known. Due to their nanometer size, the nanomaterials are already known to have many novel properties.

Many novel applications of the nanomaterials raised from these novel properties have also been proposed. Nanomaterials have size dependent properties and these can be recognized as follows:

1. Chemical properties - Reactivity, Catalysis.
2. Thermal properties - Melting temperature.
3. Mechanical properties - Adhesion, Capillary forces.
4. Optical properties - Absorption and Scattering of light.
5. Electrical properties - Tunneling of current.
6. Magnetic properties - Super paramagnetic effect.

1.6 CLASSIFICATION OF NANOSTRUCTURED MATERIALS

Nanomaterials have tremendously small size i.e. less than 100 nm and it can be in one dimension (e.g. surface films), two dimensions (e.g. fibers), or three dimensions (e.g. particles). They can exist in single, fused, aggregated or agglomerated forms with spherical, tubular, and irregular shapes. Common types of nanomaterials include nanotubes, dendrimers, quantum dots and fullerenes. Nanomaterials have applications in the field of nanotechnology, and displays different physical and chemical characteristics from normal chemicals like silver nano, carbon nanotube, fullerene, photo-catalyst, silica, etc.

Nanomaterials can be categorized as per different methods of classification including origin, dimension, and their structural configuration.

1. According to Origin of Nanomaterials:

According to origin of nanomaterials, the nanostructured materials are classified as;

(a) Natural nanomaterials: The materials belonging to the resource of nature are called natural nanomaterials. Example: Clays, fog, gelatin, etc.

(b) Artificial nanomaterials: The materials prepared through fabrication process are called artificial nanomaterials. Example: nanotubes, quantum dots, etc.

2. According to Dimensions:

According to dimensions, nanostructured materials are classified as zero dimensional (0-D), one dimensional (1-D), two dimensional (2-D), and three dimensional (3-D) nanostructures.

Zero dimensional (0-D) nanomaterials are in nano dimensions e.g. gold-silver nanoparticles. One dimensional (1-D) nanomaterials are outside nano dimensions e.g. nanowires, nanotubes. Two dimensional (2-D) nanomaterials are outside nano dimensions and include coating e.g. nanosheets and nanowalls. Three dimensional (3-D) nanomaterials are outside nano dimensions and include bulk materials e.g. multilayer films, nanofilms.

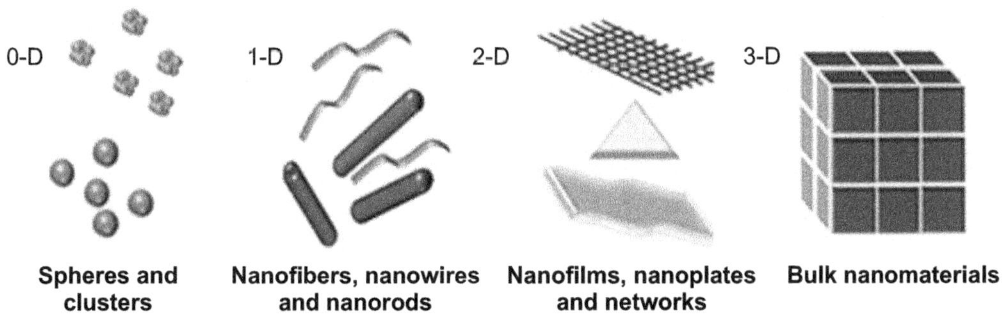

Figure 1.1: Classification of Nanomaterials

Nanomaterials are materials which are characterized by an ultra-fine grain size (< 50 nm) or by a dimensionality limited to 50 nm. Nanomaterials can be created with various modulation dimensionalities as defined by Richard W. Siegel: zero (atomic clusters, filaments and cluster assemblies), one (multilayers), and two (ultrafine-grained overlayers or buried layers), and three (nanophase materials consisting of equated nanometer sized grains) as shown in Fig. 1.1.

3. According to Structural Configuration:

According to structural configuration, nanostructured materials are classified as carbon based nanomaterials, metal based nanomaterials, dendrimers and composites. Carbon based nanomaterials may be in shape of hollow spheres, spherical, cylindrical or tubes. Metal based nanomaterials are metals while composites are multiphase solid materials and dendrimers are macromolecules.

EXERCISE

Short-Answer Questions :

1. Define nanotechnology.
2. What are the new commercial applications of nanotechnology for the next several years?
3. State any two issues that examine the challenges in nanotechnology.
4. State any two applications of nanomaterials.
5. State various types of nanomaterials.
6. What are the natural nanomaterials?
7. Give the classification of nanomaterials according to basis of structural configuration.
8. What is the significance of nanosize materials?
9. What do you mean by nano scale?
10. Give any two examples of artificial nanomaterials.

Long-Answer Questions :

1. Define nanotechnology. Give various new commercial applications for nanotechnology for the next several years.
2. Explain various issues that examine the challenges in nanotechnology.
3. Explain various challenges in nanotechnology.
4. Explain history of nanotechnology in brief.
5. What are the mile stones in evolution of nanotechnology? Explain in brief.
6. Write a short note on nanomaterials and its structure.
7. Explain the significance of nanosize materials?

☆☆☆

Chapter 2 ...

METHODS OF SYNTHESIS OF NANOMATERIALS

- Introduction
- Methods of Synthesis of Nanomaterials
- Bottom-up and Top-down Approaches
- Classification of Synthesis of Nanomaterials
- Physical Methods, Chemical methods, Hybrid methods

2.1 INTRODUCTION

Scientists are conducting research to develop novel materials with better properties, more functionality and lower cost than the existing one. Several physical, chemical, biological, and hybrid methods have been developed to enhance the performance of nanomaterials displaying improved properties with the aim. Synthesis of nanoparticles has a better control over the particle size distribution, morphology, purity, quantity, quality and environment friendly. The choice of synthesis technique can be a key factor in determining the effectiveness of studies.

2.2 METHODS OF SYNTHESIS OF NANOMATERIALS

All synthesis methods are divided into two categories based on phase of starting materials. Top-down (destruction) and bottom-up (construction) are the two main approaches for nanomaterials synthesis.

1. Bottom-up:

Bottom-up approach is very important in nanotechnology to synthesize, fabrication and processing of nanostructures and nanomaterials. A typical material synthesis is to build atom by atom on very large scale and used in industries. e.g. the deposition of films, the production of salt and nitrate in chemical industry and the growth of single crystals. Bottom-up approach denotes to the building of a nanomaterial from the bottom such as atom-by-atom or molecule-by-molecule or cluster-by-cluster till nanostructure is obtained. This route is more often used for preparing most of the nanoscale materials with the ability to generate a uniform size, shape and distribution. It effectively covers chemical synthesis and precisely controlled the reaction to inhibit further particle growth. Although the bottom-up approach is nothing new, it plays an important role in the fabrication and processing of nanostructures and nanomaterials.

For example, polymers are synthesized by connecting individual monomers together. Hence, nanostructured atoms obtained from bottom-up approach have less impurities and more homogeneity in structure.

2. Top-down:

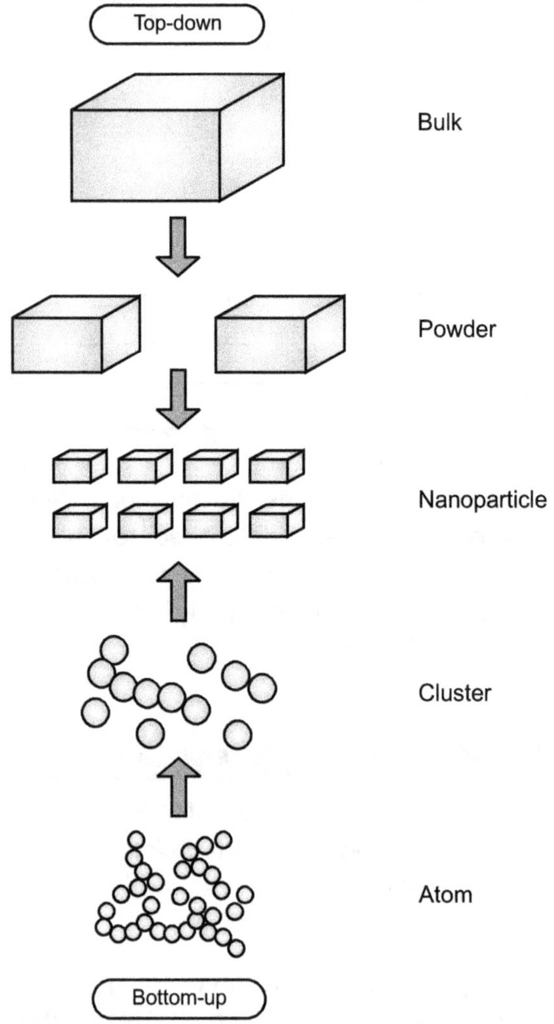

Fig. 2.1: A schematic representation of bottom-up and top-down approach

2.3 CLASSIFICATION OF SYNTHESIS METHODS OF NANOMATERIALS

Nanotechnology is an interdisciplinary subject. Scientists and researchers from many countries are working to develop materials which are stronger, lighter than existing materials and capable of use at elevated temperatures. Many processing methods have been developed to improve the performance of the existing materials. There are many techniques available to synthesize different types of nanomaterials in the form of powders, tubes, rods, wires, thin films, colloids, clusters, etc. There are various physical, chemical, biological and hybrid techniques available to synthesize the nanomaterials.

The various methods of preparation can be broadly classified under these categories:

1. **Physical:** Physical vapour deposition, Sputter deposition, Laser Ablation, Ionized cluster beam deposition, Melt mixing method, etc.
2. **Chemical:** Sol-gel method, Microwave synthesis, Hydrothermal synthesis, Colloidal route, Colloids solution, Chemical synthesis, etc.
3. **Biological:** Synthesis using DNA, synthesis using plant extracts, synthesis using micro-organisms, etc.
4. **Hybrid:** Chemical vapour deposition, Electrochemical, Electronic arc deposition, etc.

The classification of synthesis methods is shown in following Fig. 2.2.

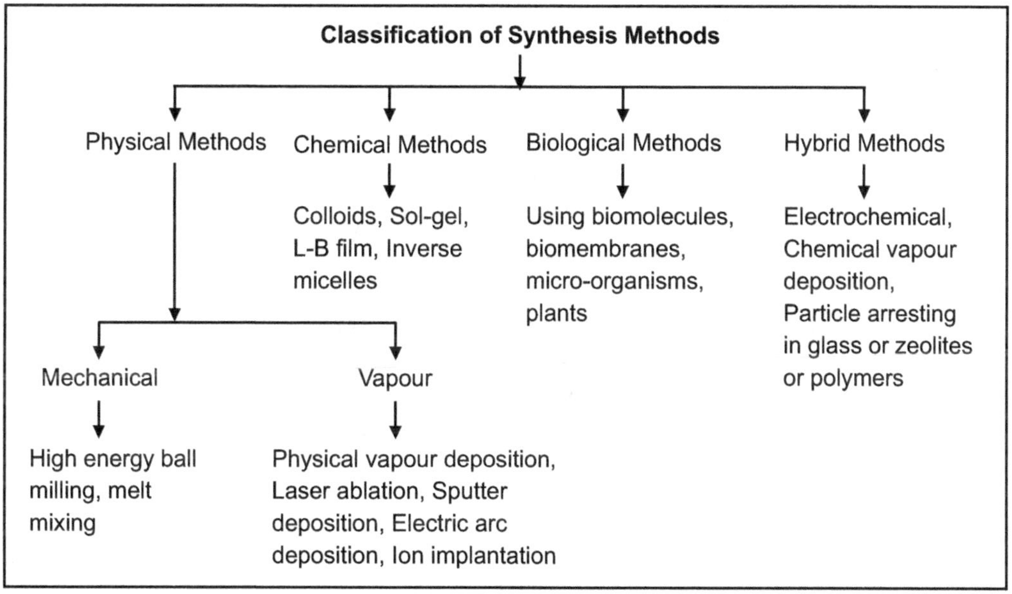

Fig. 2.2

2.3.1 Physical Methods

1. High Energy Ball Milling:

It is a ball milling process where a powder mixture placed in the ball mill is subjected to high-energy collision from the balls. This process was developed by Benjamin and his co-workers at the International Nickel Company in the late of 1960. It was found that this method, termed mechanical alloying, could successfully produce fine, uniform dispersions of oxide particles (Al_2O_3, Y_2O_3, ThO_2) in nickel-base super alloys that could not be made by more conventional powder metallurgy methods. Their innovation has changed the traditional method in which production of materials is carried out by high temperature synthesis. Besides materials synthesis, high-energy ball milling is a way of modifying the conditions in which chemical reactions usually take place either by changing the reactivity of as-milled solids (mechanical activation - increasing reaction rates, lowering reaction temperature of the ground powders)- or by inducing chemical reactions during milling

(Chemo-mechanic). It is furthermore, a way of inducing phase transformations in starting powders whose particles have all the same chemical composition: amorphization or polymorphic transformations of compounds, disordering of ordered alloys, etc.

Fig. 2.3 shows the motions of the balls and the powder. Since the rotation directions of the bowl and turn disc are opposite, the centrifugal forces are alternately synchronized. Thus friction resulted from the hardened milling balls and the powder mixture being ground alternately rolling on the inner wall of the bowl and striking the opposite wall. The impact energy of the milling balls in the normal direction attains a value of up to 40 times higher than that due to gravitational acceleration. Hence, the planetary ball mill can be used for high-speed milling.

Schematic view of motion of the ball and powder mixture is shown in Fig. 2.3.

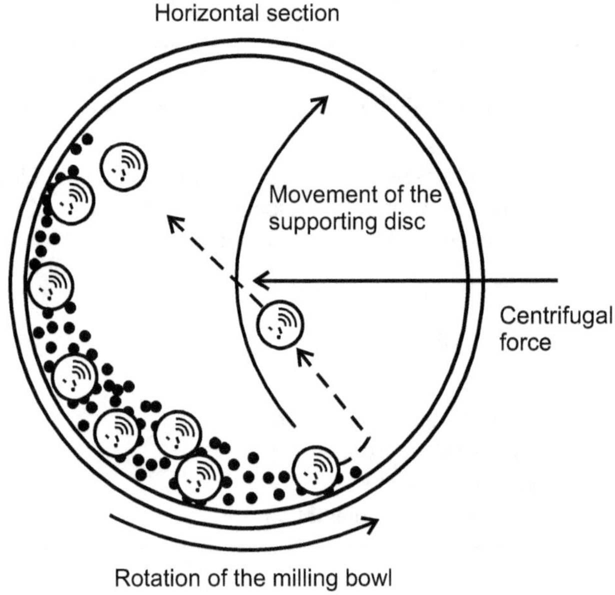

Fig. 2.3: Schematic view of motion of the ball and powder mixture

During the high-energy ball milling process, the powder particles are subjected to high energetic impact. Micro-structurally, the mechanical alloying process can be divided into four stages: (a) initial stage, (b) intermediate stage, (c) final stage, and (d) completion stage.

(a) At the initial stage of ball milling, the powder particles are flattened by the compressive forces due to the collision of the balls. Micro-forging leads to changes in the shapes of individual particles, or cluster of particles being impacted repeatedly by the milling balls with high kinetic energy. However, such deformation of the powders shows no net change in mass.

(b) At the intermediate stage of the mechanical alloying process, significant changes occur in comparison with those in the initial stage. Cold welding is now significant. The intimate mixture of the powder constituents decreases the diffusion distance to the micrometer range. Fracturing and cold welding are the dominant milling processes at this stage. Although some dissolution may take place, the chemical composition of the alloyed powder is still not homogeneous.

(c) At the final stage of the mechanical alloying process, considerable refinement and reduction in particle size is evident. The microstructure of the particle also appears to be more homogeneous in microscopic scale than those at the initial and intermediate stages. True alloys may have already been formed.

(d) At the completion stage of the mechanical alloying process, the powder particles possess an extremely deformed metastable structure. At this stage, the lamellae are no longer resolvable by optical microscopy. Further mechanical alloying beyond this stage cannot physically improve the dispersoid distribution. Real alloy with composition similar to the starting constituents is thus formed.

2. Physical Vapour Deposition:

Physical Vapour Deposition (PVD) is fundamentally a vaporization coating technique, involving transfer of material on an atomic level. It is an alternative process to electroplating. Physical vapour deposition is a collective set of processes used to deposit thin layers of material, typically in the range of few nanometers to several micrometers. Physical vapour deposition techniques consist of four fundamental steps:

- Evaporation
- Transportation
- Reaction
- Deposition

The two most common PVD processes are thermal evaporation and sputtering. Thermal evaporation is a deposition technique that relies on vaporization of source material by heating the material using appropriate methods in vacuum. Sputtering is a plasma-assisted technique that creates a vapour from the source target through bombardment with accelerated gaseous ions (typically Argon). In both evaporation and sputtering, the resulting vapour phase is subsequently deposited onto the desired substrate through a condensation mechanism.

(a) Evaporation: During this stage, a target, consisting of the material to be deposited is bombarded by a high energy source such as a beam of electrons or ions. This dislodges atoms from the surface of the target, 'vaporizing' them.

(b) Transport: This process simply consists of the movement of 'vaporized' atoms from the target to the substrate to be coated and will generally be a straight line affair.

(c) Reaction: In some cases coatings will consist of metal oxides, nitrides, carbides and other such materials. In these cases, the target will consist of the metal. The atoms of metal will then react with the appropriate gas during the transport stage. For the above examples, the reactive gases may be oxygen, nitrogen and methane.

In instances where the coating consists of the target material alone, this step would not be part of the process.

(d) Deposition: This is the process of coating build up on the substrate surface.

Depending on the actual process, some reactions between target materials and the reactive gases may also take place at the substrate surface simultaneously with the deposition process.

PVD coatings are deposited for numerous reasons. Some of the main ones are:

- Improved hardness and wear resistance
- Reduced friction
- Improved oxidation resistance

The use of such coatings is aimed at improving efficiency through improved performance and longer component life. They may also allow coated components to operate in environments that the uncoated component would not otherwise have been able to perform.

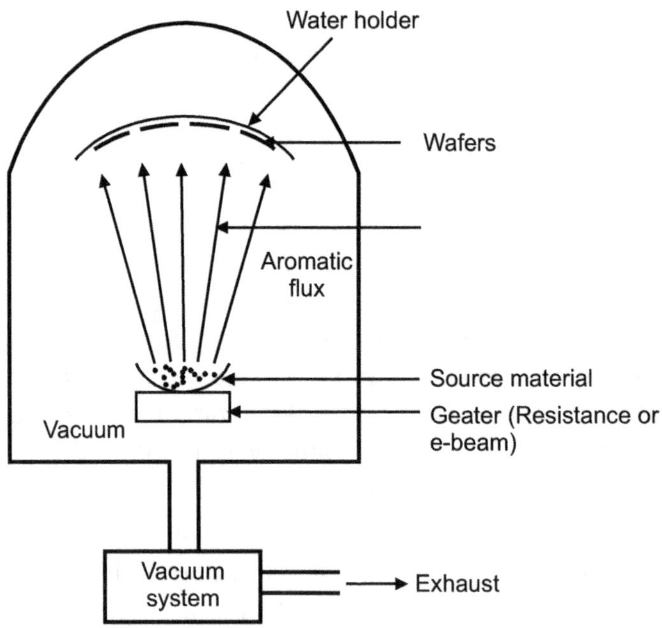

Fig. 2.4: Schematic representation of Physical Vapour Deposition (PVD)

Advantages:

- Materials can be deposited with improved properties compared to the substrate material.
- Almost any type of inorganic material can be used as well as some kinds of organic materials.
- The process is more environmental friendly than processes such as electroplating.

Disadvantages:

- It is a line of sight technique meaning that it is extremely difficult to coat undercuts and similar surface features.
- High capital cost.
- Some processes operate at high vacuums and temperatures requiring skilled operators.
- Processes requiring large amounts of heat require appropriate cooling systems.
- The rate of coating deposition is usually quite slow.

Applications:

PVD coatings are generally used to improve hardness, wear resistance and oxidation resistance. Thus, such coatings are used in a wide range of applications such as:

- Aerospace
- Automotive
- Surgical/Medical
- Dies and moulds for all manner of material processing
- Cutting tools
- Fire arms

3. Thermal Evaporation:

Thermal evaporation is one of the simplest and most popular synthesis physical vapour deposition methods. It has been very successful and versatile in fabricating nanobelts and nanowires with various characteristics. The basic process of this method is sublimating source material(s) in powder form at high temperature, and a subsequent deposition of the vapour in a certain temperature region to form desired nanostructures.

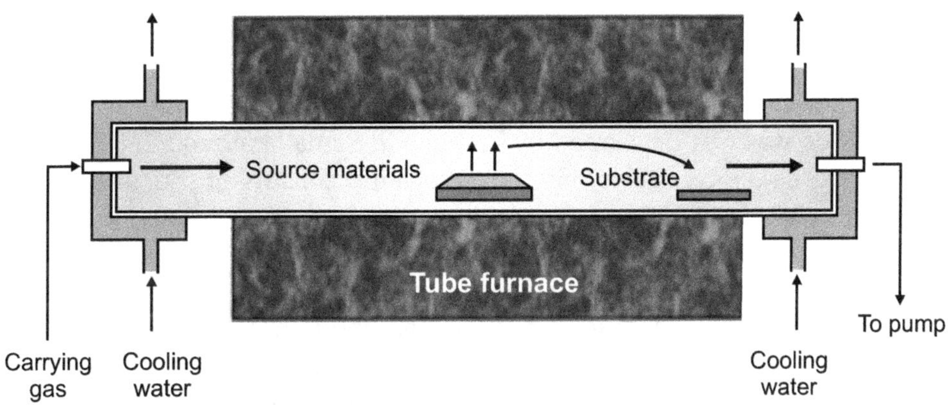

Fig. 2.5: Thermal evaporate deposition system for synthesis of 1-D nanostructures

A typical experimental system is shown in Fig. 2.5. The synthesis is performed in an alumina or quartz tube, which is located in a horizontal tube furnace. High purity oxide powders contained in an alumina boat are loaded in the middle of the furnace, the highest temperature region. The substrates for collecting the desired nanostructures are usually placed down-stream following the carrier gas. The substrates can be silicon wafer, poly-crystalline alumina or single crystal alumina (sapphire). Both ends of the tube are covered by stainless steel caps and sealed with O-rings. Cooling water flows inside the cover caps to achieve a reasonable temperature gradient in the tube.

During the experiments, the system is first pumped down to around 2-10 Torr. Then the furnace is turned on to heat the tube to the reaction temperature at a specific heating rate. An inert carrying gas, such as argon or nitrogen, is then introduced into the system at a constant flow rate to bring the pressure in the tube back to 200-500 Torr (different pressures are required by different source materials and final deposited nanostructures). The reaction temperature and pressure are held for a certain period of time to vaporize the source material and achieve a reasonable amount of deposition.

Source materials can be vaporized at the high temperature and low pressure condition. The vapor is then carried by the inert carrying gas down to the lower temperature region, where the vapor gradually becomes supersaturated. Once it reaches the substrate, nucleation and growth of nanostructures will occur. The growth is terminated when the furnace is turned off. The system is then cooled down to room temperature with flowing inert gas.

The thermal evaporation process is basically a physical vapor deposition process and has been successfully used for synthesizing a variety of oxide and non-oxide nanobelts and nanowires. Moreover, this system can also be used for chemical vapor deposition (CVD) by simply applying reaction gases instead of the carrier gas and placing substrates in the middle of the tube. For example, multi-wall and single-wall carbon nanotubes have been successfully fabricated in this system using hydrogen and methane/acetylene as reactants. Metal catalysts, such as gold, tin, copper, etc. have also been used to achieve size control and alignment.

4. Ionized Cluster Beam Deposition:

Thermal evaporation is a physical vapour deposition method. ICBD (Ionized cluster beam deposition) method was first developed by Takagi and Yamada around 1985. It is useful to obtain high quality single crystalline thin film. The setup consists of source of evaporation, a nozzle through which material can expand into the chambers, an electron beam ionized the clusters, an arrangement to accelerate the clusters and a substrate on which nanoparticle film can be deposited. A schematic arrangement is shown in following figure. Small clusters from molten material are expanded through a fine nozzle. The vapour pressure of 1 Pa to 1 KPa needs to be created into sources. The nozzle needs to have a diameter larger than mean free path of molecule in vapour form in the source to form a cluster. On collision with electron beam, cluster gets ionized. Due to applied accelerating voltage, clusters are directed towards substrate and by controlling accelerated voltage, it is possible to control energy which cluster hits the substrate. Stable clusters of some materials require energy to break their bonding structure and produce small clusters.

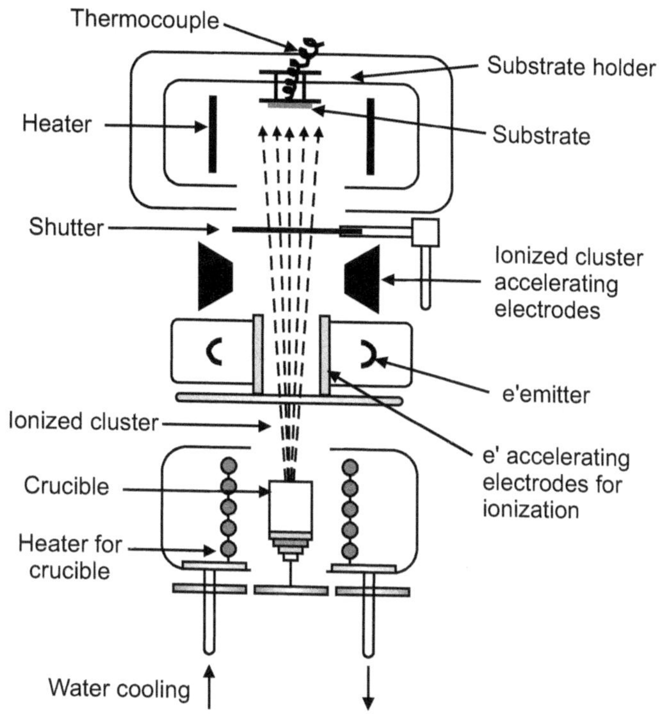

Fig. 2.6: Schematic diagram of ionized cluster beam deposition

5. Sputtering Deposition:

Sputter deposition is a widely used technique to deposit thin films on substrates. The technique is based upon ion bombardment of a source material, the target. Ion bombardment results in a vapor due to a purely physical process, i.e. the sputtering of the target material. Hence, this technique is part of the class of physical vapor deposition techniques which includes, for example, thermal evaporation and pulsed laser deposition. The most common approach for growing thin films by sputter deposition is the use of a magnetron source in which positive ions present in the plasma of a magnetically enhanced glow discharge bombard the target. The target can be powered in different ways, ranging from dc for conductive targets, to RF for nonconductive targets, to a variety of different ways of applying current and/or voltage pulses to the target. Sputtering is a purely physical process. For example, deposit of compound layer must be done adhoc through the addition of a reactive gas (Ar) to the plasma, i.e. reactive sputtering. The undesirable reaction of the reactive gas with the target material results in a non-linear behavior of the deposition parameters as a function of the reactive gas (Ar gas) flow. To model this behaviour, the fluxes of the various species towards the target must be determined. However, equally important are the fluxes of species incident at the substrate because they not only influence the reactive sputter deposition process, but also control the growth of the desired film.

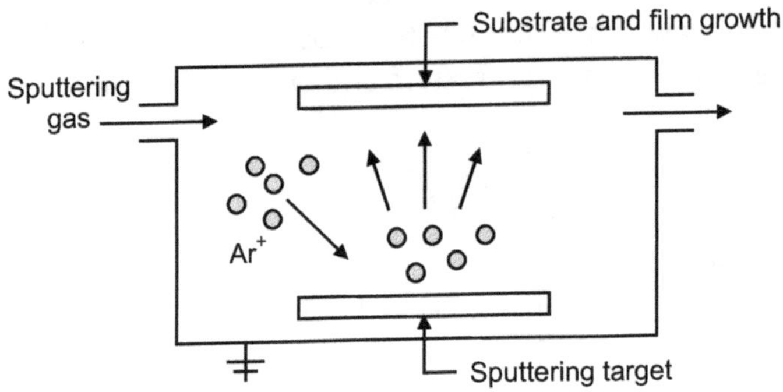

Fig. 2.7 (a): Schematic diagram of Sputter deposition

The sputtering method of thin film deposition involves introducing a controlled gas, usually chemically inert argon, into a vacuum chamber, and electrically energizing a cathode to establish self-sustaining plasma. The exposed surface of the cathode is called the target viz. a slab of the material to be coated onto the substrates.

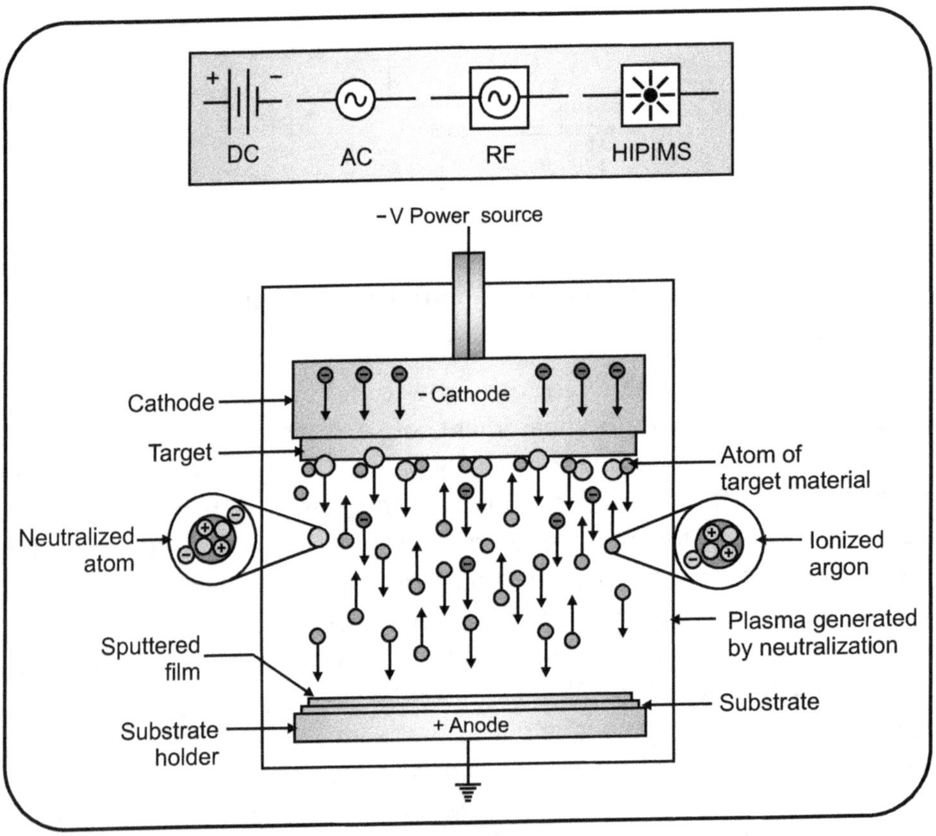

Fig. 2.7 (b): Schematic representation of sputter process

The gas atoms lose electrons inside the plasma to become positively charged ions, which are then accelerated into the target and strike with sufficient kinetic energy to dislodge atoms or molecules of the target material. It can be thought of as a sort of atomic scale bead blasting. This sputtered material now constitutes a vapour stream, which traverses the chamber and hits the substrate, sticking to it as a coating or "thin film".

In order to get good film adhesion, it is, of course, necessary for the substrate surface to be clean. Appropriate cleaning and handling steps must be employed prior to placing substrates into the vacuum chamber. In addition, it is not uncommon to also have optional in situ cleaning features such as sputter etch incorporated into the sputter system.

Steps may have to be taken to control various film properties as desired. Fortunately, sputter system design can allow adjustability of a number of parameters in order to give process engineers the ability to achieve desired results for such variables as thickness, uniformity, adhesion strength, stress, grain structure, optical or electrical properties, etc.

There are also considerations such as what type of power to use on the cathodes. DC power is suitable for conductive materials, but RF power can also sputter non-conductive materials. Pulsed DC has advantages for some processes such as reactive sputtering.

6. Ultrasonic Spray Pyrolysis:

Spray pyrolysis is one of the most cost effective techniques with simple apparatus for deposition of uniform thin films of large surface areas under atmospheric conditions. It is performed at a relatively low temperature and at a high deposition rate. It has the capability of controlling the shape, size, composition and phase homogeneity of the particles due to the easy control of parameters. The extremely large choice of precursors along with simple equipment for mass production of large areas have made spray pyrolysis an industry compatible manufacturing technique for synthesis of porous films with well-controlled microstructures. It has a great potential for producing functionally graded electrodes that require deposition of multiple layers of material, each with a different pore structure.

Ultrasonic Spray Pyrolysis (USP) is one of the important techniques to synthesize nano-materials and to coat substrates with thin film coatings. Ultrasonic spraying and atomization allows for full process control resulting in homogeneous high-quality output. Ultrasonic spray pyrolysis excels conventional techniques, e.g. CVD by its homogeneous distribution and its cost-efficiency. Ultrasonic Spray Pyrolysis (USP) is a simple aerosol synthetic technique widely used for the synthesis of nanomaterials such as thin films or nanoparticles. Due to its easy feasibility, flexibility and cost-efficiency, the USP method is an important alternative to the chemical vapour deposition (CVD). The precursors for ultrasonic spray pyrolysis are often prepared via sol-gel route. The composition of the synthesized nanoparticles or the film can be easily modified via changes of the processing parameters.

Ultrasonic spray pyrolysis gives you full control over the most important process parameters such as:

- ultrasonic amplitude
- precursor solution
- precursor composition/ viscosity
- flow rate
- deposition temperature
- substrate temperature

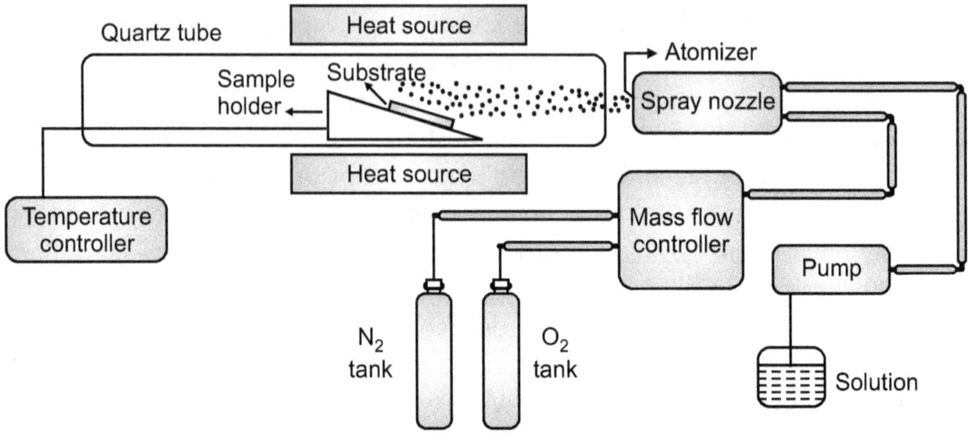

Fig. 2.8: Schematic of the spray pyrolysis setup

(i) Ultrasonic Spray Pyrolysis for the Synthesis of Nanoparticles: When nanoparticles are synthesized via ultrasonic spray pyrolysis, organic and inorganic salts can be used as precursors for the preparation of metallic, oxidic and composite nanopowders. The precursor solution is ultrasonically atomized before the fine-sized droplet mist goes into the furnace, where thermal decomposition of the material takes place. USP synthesized powders can be divided in three groups:

- Metals: Ag, Co, Cu, Zn, Ni, Fe,etc.
- Oxides: TiO_2, ZnO, Al_2O_3, etc.
- Composite: CuNi, FeCo, NiCo, etc.

(ii) Ultrasonic Spray Pyrolysis for Thin Film Coatings: Via ultrasonic spray pyrolysis multi-layer coating can be produced, which is of special interest for the preparation of functionally graded films. By changing the precursor solution, the composition and the functionalities of the thin film coating can be easily and precisely modified.

Advantages of Ultrasonic Spray Pyrolysis:

- Cost effective method due to high transfer efficiency in thin film formation.
- Substrates with complex geometries can be spray coated.
- Films deposited by ultrasonic spray pyrolysis are reproducible.
- Highly uniform nanoparticle size formation.
- Non-clogging nozzles and robust design.

2.3.2 Chemical Methods

1. Colloidal Method:

Colloidal methods are simple and well established wet chemistry precipitation processes in which solutions of the different ions are mixed under controlled temperature and pressure to form insoluble precipitates. For metal nanoparticles, the basic principles of colloidal preparation were known since antiquity. E.g. gold colloids used for high quality red and purple stained glass from medieval times to date. However, proper scientific investigations of colloidal preparation methods started only in 1857 when Faraday has published results of his experiments with gold. He prepared gold colloids by reduction of $HAuCl_4$ with phosphorus. Today, colloidal processes are widely used to produce such nanomaterials like metals, metal oxides, organics, and pharmaceuticals. A typical chemical reactor used to synthesize nanoparticles using colloidal method is shown in Fig. 2.9.

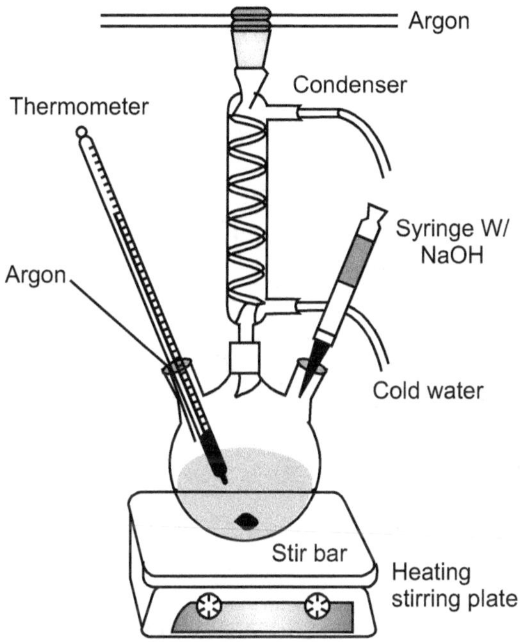

Fig. 2.9: Synthesis of nanoparticles by colloidal method

Chemical reactions in which colloidal particles are obtained are carried out in some glass reactor of suitable size. Glass reactor usually has a provision to introduce some precursors, gases as well as to measure temperature, pH, etc. during the reaction. It is usually possible to remove the products at suitable time intervals. Reaction is usually carried out under inert atmosphere like Ar or N_2 gas so as to avoid any uncontrolled oxidation of the products. There is also a provision made to stir the reactants during the reaction by using Teflon coated magnetic needle.

2. Co-precipitation Method:

Crystalline precipitates separating from a solution are not quite pure, but usually contain imperfection of mother liquor with foreign constituents. A vast number of empirical facts pertaining to the presence of various impurities in crystalline precipitates may be found is called co-precipitation or carrying down. The schematic representation of co-precipitation for nanoparticles is shown in Fig. 2.10. The experimental setup consists of glass round bottom flask placed on heating mantle. The deionized distilled water is taken in the round bottom flask. After reaching thermal equilibrium, the weighed amount of precursor is added into the flask. Then complex agent like ammonium hydroxide is added in the flask. The solution is stirring. The precipitation starts and allowed for cooling and setting of the precipitation in the bottom of the flask. After precipitation, the supernatant liquid is first decanted carefully into filter paper without disturbing the precipitate. The filtrate is taken into centrifuged tubes and after centrifugation, supernatant liquid is allowed to pass through the filter paper followed by the transfer of the precipitate. Finally, using ethanol solution, the precipitate is washed and dried at high temperature in vacuum to obtain nanoparticles.

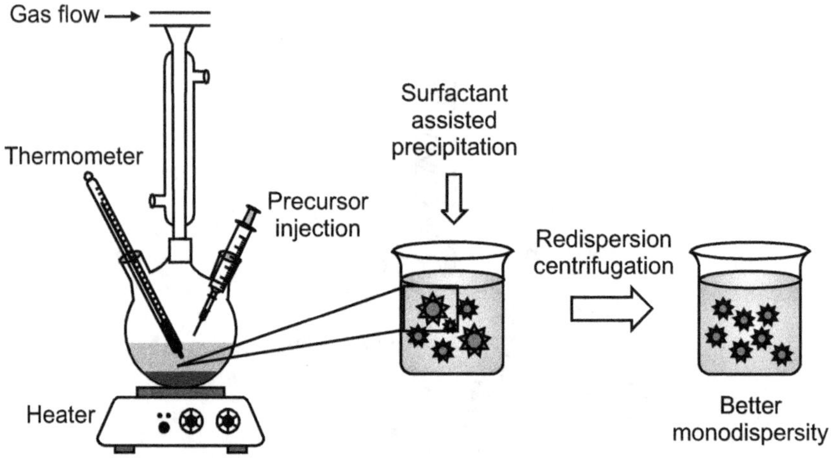

Fig. 2.10: Schematic representation of co-precipitation for nanoparticles

3. Sol-gel Method:

Sol-gel process consists of the chemical transformation of a sol (liquid) into a gel state and with subsequent post-treatment and transition into solid oxide material. The sol-gel method was developed in the 1960s mainly due to the need of new synthesis methods in the nuclear industry. A method is needed to lower sintering temperature. The idea behind sol-gel synthesis is to "dissolve" the compound in a liquid in order to bring it back as a solid in a controlled manner. A sol is a stable dispersion of colloidal particles in a solvent and a gel consists of a three dimensional continuous network, which encloses a liquid phase. In a colloidal gel, the network is built from agglomeration of colloidal particles. Multicomponent compounds may be prepared with a controlled stoichiometry by mixing sols of different compounds. The sol-gel method prevents the problems with co-precipitation, which may be inhomogeneous and a gelation reaction. It enables to mixing at an atomic level and results in small particles, which are easily sinterable.

Sol-gel synthesis may be used to prepare materials with a variety of shapes, such as porous structures, thin fibers, dense powders and thin films. If the gel is dried by evaporation, then the capillary forces will result in shrinkage, the gel network will collapse, and a xerogel is formed. If drying is performed under supercritical conditions, the network structure may be retained and a gel with large pores may be formed. This is called an aerogel, and the density will be very low i.e. < 0.005 g/cm^3. The main benefits of sol-gel processing are the high purity and uniform nanostructure achievable at low temperatures.

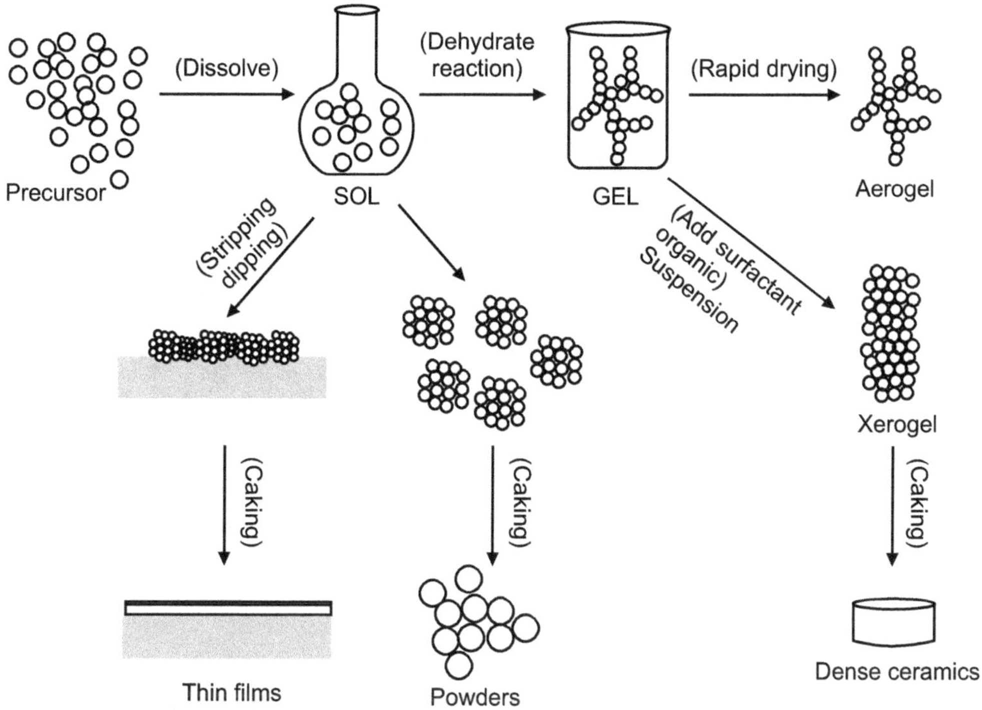

Fig. 2.11: Different stages of synthesis of nanoparticles using sol-gel method

2.3.3 Hybrid Methods

1. Electrochemical Method:

Electrochemical deposition is a technique that has been widely used for the synthesis of metal nanoparticles. Electrochemical deposition occurs at the interface of an electrolyte solution containing the metal to be deposited and an electrically conductive metal substrate. Electrochemical deposition is a very attractive method because the process is simple and effective. The principle of electro-deposition is inducing chemical reactions in an aqueous electrolyte solution with the help of applied voltage. e.g. this is the process of using electrical current to coat an electrically conductive object with a relatively thin layer of metal. This method is relevant to deposition of nanostructured materials. There is a huge variety of possibly reduced ions, and it has no limitation as far as sample shape and size are concerned. Deposition of the layer can be done both in constant, also called direct current (DC), and accelerating current (AC) modes depending on the needs, possibilities or applications and wires characteristics. In the first case, constant current is applied to electrodes whereas in the second one the potential of working electrode is controlled. Electrochemical deposition of nanowires is a technique which combines either bottom-up or top-down approaches. This is due to the fact that the wires have grown atom by atom manner and it can be obtained in the matrices which were subjected to anodization process during which the nano-pores are obtained in bulk material.

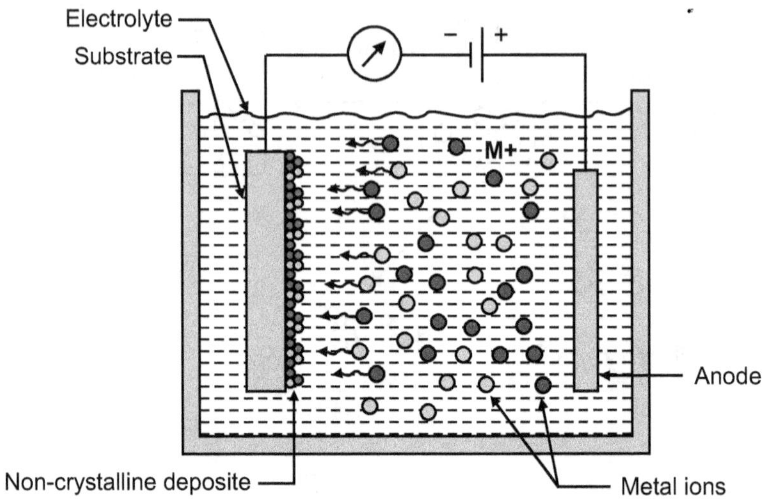

Fig. 2.12: Schematic diagram for electrochemical deposition method

Before the templates can be used for the deposition;

- it should be chemically inactive in a particular process;
- deposited material should wet the pores wall;
- deposition process should not be chaotic and start from the bottom or from the wall of the pore and go upwards or to the centre, respectively;

Advantages:

- The process is relatively cheap and fast and allows complex shapes.
- The thickness of layer simply depends on the current density and the time for which the current flows.
- The deposit can be detached if the substrate is chosen to be soluble by dissolving it away.
- Electro-deposition has three main attributes that make it so well suited for nano, bio and micro technologies.
- It can be used to grow functional material through complex 3D masks.
- It can be performed near room temperature from water-based electrolytes.
- It can be scaled down to the deposition of a few atoms or up to large dimensions.

2. Chemical Vapour Deposition:

Chemical vapour deposition (CVD) involves a chemical reaction followed by the deposition used to produce high quality and high performance solid materials. Generally, this process is used in the semiconductor industry to produce thin films.

The basic principle of CVD is the reactant particles are mixed with carrier gas and allowed to pass on through the hot solid substrate surface. When the reactant particles and substrate comes in contact, the heat energy initiates the chemical reactions and form nanofilm on the substrate surface. The byproducts of the reactions on the solid surface can be removed by washing. These precursors decompose on the substrate and produce the desired deposit. In CVD, the vaporized precursors are introduced into a CVD reactor and adsorb onto a substance held at an elevated temperature. In CVD process, vaporized precursors are made to adsorb onto a substance held at a high temperature. The adsorbed molecules react with other or decompose and produce crystals.

There are three main steps in CVD process:

1. Reactants are transported on the growth surface by a boundary layer.
2. Chemical reactions take place on the growth surface.
3. By products formed by the gas-phase reactions have to be removed from the surface.

Homogeneous nucleation takes place in gas phase. Heterogeneous nucleation takes place in substrate. Schematic diagram of the chemical vapor deposition apparatus is shown in Fig. 2.13.

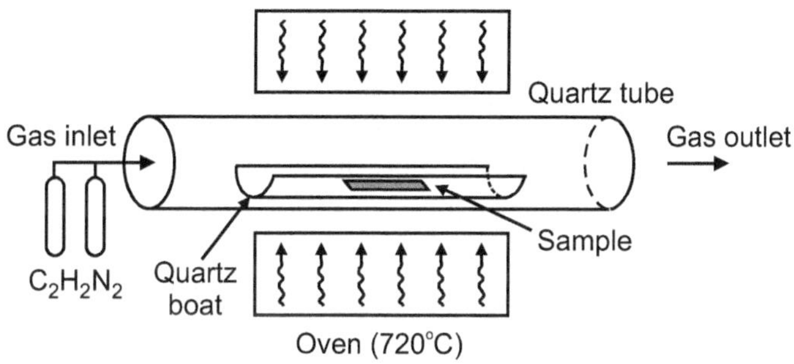

Fig. 2.13: Typical diagram for chemical vapour deposition

CVD involves the flow of gas with diffused reactants (substances to be deposited in vapour) over a hot substrate surface. The gas that carries the reactants is called the carrier gas. While the gas flows over the hot solid surface, the heat energy increases chemical reactions of the reactants that form film during and after the reactions. The byproduct of reactions is removed to get the thin film of desired composition.

Advantages:

- To produce defect free nanoparticles.
- Scaling up of the CVD system for manufacturing purpose is achieved without any major difficulties.

EXERCISE

Short-Answer Questions:

1. What is Bottom-up approach?
2. What is Top-down approach?
3. State various physical methods of synthesis of nanomaterials.
4. State various chemical methods of synthesis of nanomaterials.
5. State various hybrid methods of synthesis of nanomaterials.

6. What are advantages of electrochemical method?
7. What are advantages of chemical vapour deposition?
8. What is the main benefit of sol-gel processing?
9. State various important parameters of ultrasonic spray pyrolysis techniques.
10. What are advantages of physical vapour deposition?
11. What are advantages of ultrasonic spray pyrolysis techniques?
12. What are disadvantages of physical vapour deposition?
13. State various applications of physical vapour deposition.
14. Draw typical diagram for:
    (a) Top-down and Bottom-up approach
    (b) Physical vapour deposition
    (c) Sputtering deposition
    (d) Colloidal method
    (e) Chemical vapour deposition

Long-Answer Questions:

1. What are the methods of synthesis of nanomaterials? Explain Top-down and Bottom-up approach with neat diagram.
2. Give the classification of synthesis methods of nanomaterials? State their various types with pictorial representation.
3. Describe synthesis of nanomaterials using high energy ball milling with suitable diagram.
4. Describe synthesis of nanomaterials using physical vapour deposition with suitable diagram.
5. Describe synthesis of nanomaterials using ionized cluster beam deposition with suitable diagram.
6. Describe synthesis of nanomaterials using high energy ball milling with suitable diagram.
7. Describe synthesis of nanomaterials using sputtering deposition with suitable diagram.
8. Describe synthesis of nanomaterials using ultrasonic spray pyrolysis with suitable diagram.
9. Describe synthesis of nanomaterials using colloidal method with suitable diagram.
10. Describe synthesis of nanomaterials using co-precipitation with suitable diagram.
11. Describe synthesis of nanomaterials using sol-gel method with suitable diagram.
12. Describe synthesis of nanomaterials using electrochemical method with suitable diagram.
13. Describe synthesis of nanomaterials using chemical vapour deposition with suitable diagram.

☆☆☆

Chapter 3 ...

CHARACTERIZATION TECHNIQUES

- Introduction
- Optical Properties: UV- Visible Spectroscopy
- X-Ray Diffraction (XRD)
- Scanning Electron Microscopy (SEM)
- Transmission Electron Microscopy (TEM)

3.1 INTRODUCTION

The information about the chemical composition, content of impurity, imperfections, etc. of materials is necessary to understand the properties of material like structural, magnetic, electrical, chemical, thermal, and optical. Researchers have developed and found many applications of nanomaterials using various synthesis techniques in the area of characterization of materials and nanotechnology. This chapter discusses various characterization techniques employed for synthesis. This chapter contains details of basic principle, working and applications of UV-Visible Spectroscopy, XRD, SEM, and TEM.

3.2 OPTICAL STUDY

3.2.1 UV-Visible Spectroscopy Technique

UV-visible spectroscopy is a spectrochemical method to characterize the absorption spectra of all atoms. Scientifically, the measurement and interpretation of electromagnetic radiations absorbed, scattered or emitted by atoms or molecules or ions is called spectroscopy. UV-visible spectroscopy is used for optical characterization of prepared samples in absorption mode.

Principle: The emission of electrons from a material surface is observed when it is illuminated by light or any other radiation of suitable wavelength.

When a beam of electromagnetic radiations is incident on sample, it can be absorbed, transmitted, scattered, reflected or it can excite fluorescence. The absorption and transmission processes are concerned in the absorption spectroscopy. UV-visible spectroscopy is used for quantitative analysis of the prepared samples or metal nanoparticles. The analytical applications of the absorption of radiation by some chemical species can be either quantitative or qualitative. The absorption spectrum provides the measurement of energy absorbed when electrons jump to higher energy levels. This transition of electrons from one fixed lower energy level to other higher energy level gives emission or absorption of radiation at discrete energies. The absorption of UV-visible radiations provides information of electronic absorption bands of many closely packed and discrete lines of prepared sample molecules and each line in the spectrum arises due to the transition of an electron from ground state to one of many vibrational or higher or excited electronic energy states. In that ultraviolet molecules absorb more energy for excitation than in the visible region.

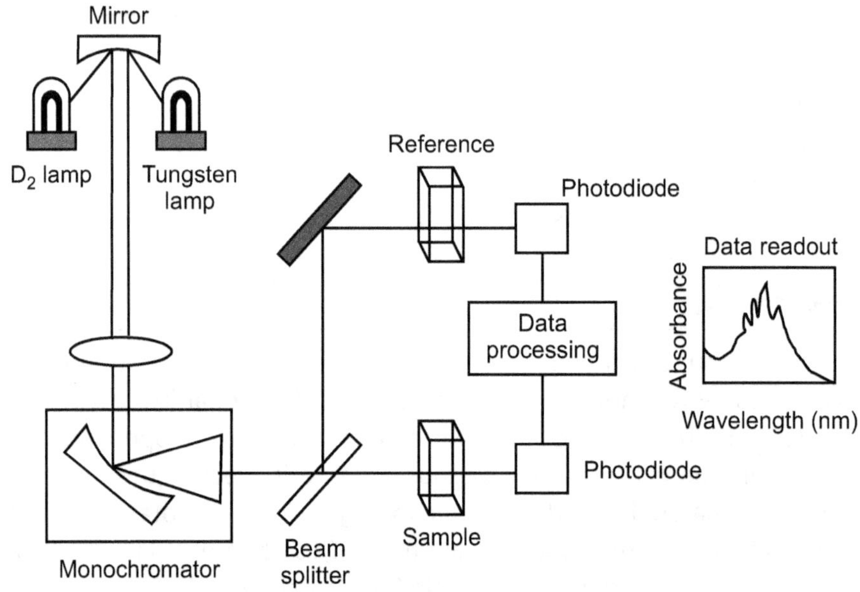

Fig. 3.1: Schematic diagram of UV-visible spectrometer

A schematic diagram of UV-visible spectrometer is shown in Fig. 3.1. The basic parts of a UV-visible spectrometer are radiation source (Tungsten filament or deuterium arc lamp or xenon arc lamp or LED), holder for the sample, diffraction grating, monochromatic prism, and detector. The detector may be any kind of photomultiplier tube or photodiode. This UV-visible optical spectrometer measures the wavelengths at which absorption occurs together with the degree of absorption at each wavelength following Beer-Lambert law. According to Beer-Lambert law, the concentration of a substance in sol is directly proportional to the absorbance.

1. The UV-Visible spectrometer records the intensity of light after transmitted (I) through the sample and the intensity of light before transmitted (I_0) through the sample. Therefore, Transmittance (T) is the ratio of intensity of light after transmitted (I) to intensity of light before transmitted (I_0) through the sample. Hence,

$$\text{Transmittance} = \frac{I}{I_0}$$

2. The absorbance (A) in terms of transmittance is,

$$A = -\log\left(\frac{T}{100}\right)$$

3. Reflectance (R) is the ratio of intensity of light after reflected (I) to intensity of light before reflected (I_0) through the sample. Hence,

$$\text{Reflectance} = \frac{I}{I_0}$$

Applications of UV-visible Spectroscopy:

Following are the some applications of UV-visible spectroscopy.

1. **Detection of impurities:** UV-Visible absorption spectroscopy is used to determine the impurities in organic molecules.
2. **Structure elucidation of organic compounds:** UV-visible absorption spectroscopy is used in the structure elucidation of organic molecules to determine the presence or absence of unsaturation and hetero atoms.
3. **Quantitative analysis:** UV-visible absorption spectroscopy is used for the quantitative determination of compounds that absorb UV-radiation using Beer-Lambert's law :

$$A = \log\left(\frac{I_0}{I_t}\right) = \log\left(\frac{1}{T}\right) = -\log T = \varepsilon bC$$

where, 'T' is extinction coefficient, ε is absorption coefficient, 'C' is concentration, and 'b' is the length of the cell used in UV-spectrophotometer.

4. **Qualitative analysis:** UV-absorption spectroscopy is used for the qualitative determination of compounds which absorb UV- radiation.

3.3 STRUCTURAL STUDY

3.3.1 X-ray Diffraction Technique (XRD)

X-ray diffraction method is analytical technique for characterizing materials which provides structural property with information about the crystallographic structure, chemical configuration and physical properties of materials. Properties of material are highly dependent on its structure, therefore phase identification is important. X-ray diffraction technique is mostly applicable to determine the mean size of nanoparticles, to determine the structure and to estimate the mean size variations with the increase in annealing temperature of nanoparticles.

Principle: The working of X-ray diffraction process is based on observing the scattered intensity of an X-ray beam hitting nanoparticles as a function of incident, scattered angle, polarization, wavelength and energy.

More number of electrons are emitted from the heated filament when current is applied to the filament within the picture tube. The electrons are accelerated due to the high 15 to 60 kV voltage applied to the picture tube to hit a target of copper and X-rays are produced with wavelength. This wavelength characterized X-rays of target are collimated and directed onto the sample which has been ground to a fine powder of nanoparticles to produce particle sizes of less than 10 microns and detects the X-ray signal using detector. Then, X-ray signals are processed and converted into count rate by microprocessor and X-ray scanner scans the rate by changing the angle between the source and sample. A ray diagram of X-Ray diffraction techniques is shown in Fig. 3.2.

The lattice constant is calculated using the relation,

$$a = d\sqrt{(h^2 + k^2 + l^2)}$$

where, a is the lattice constant, d is the inter-planar spacing, and (h, k and l) are the Miller indices.

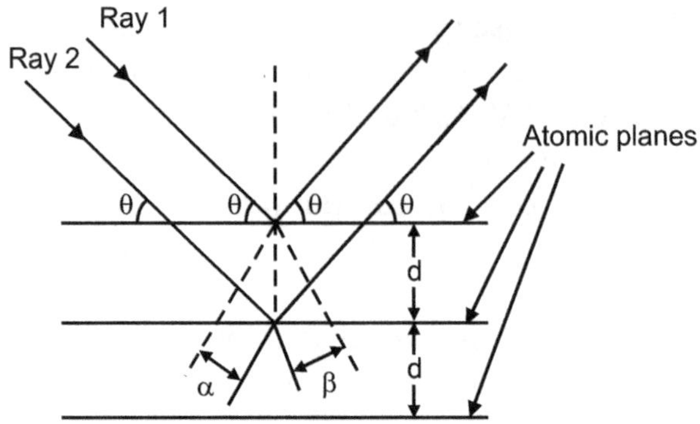

Fig. 3.2: X-Ray diffraction techniques

According to Bragg's law, by hitting X-ray beam on prepared sample, the spacing between the planes of the atoms that constitute the sample can be measured and it is diffracted.

$$N\lambda = 2d \sin \theta$$

where, n is the order of the diffracted beam,

λ is the wavelength of the incident X-ray beam,

d is the spacing between adjacent planes of atoms, and

θ is the angle of incidence of the X-ray beam.

The XRD data were used to estimate mean particle size (t) of nanoparticles in the materials by using Scherer formula;

$$T = \frac{0.9 \lambda}{\beta \cos \theta}$$

where, λ is the wavelength of X-ray radiation,

and β is an angular full width at half maximum of the X-ray diffraction peak at the diffraction angle θ.

Advantages of X-Ray Diffraction:

- XRD units are widely available.
- It requires minimum sample preparation.
- It is a powerful and faster (< 20 min) technique for identification of an unknown mineral.
- It provides an unambiguous mineral determination.
- XRD data interpretation is relatively straight forward.

Disadvantages of X-Ray Diffraction:

- Homogeneous and single phase material is best for identification.
- It refers standard reference file (JCPD) of inorganic compounds for d-spacing and (h, k, *l*) planes.
- Detection limit is only 2% for mixed materials.
- XRD patterns indexing for non-isometric crystal systems are complicated.
- For high angle reflections, peak overlay may occur.

Applications of X-Ray Diffraction Technique:

Following are the few applications of X-ray diffraction technique.

- To analyse the complete structure of crystalline materials.
- To determine phase transitions in a given substance.
- To measure the crystal size (grain size) particularly in nanoscale of materials.
- To investigate stress and preferred orientation in polycrystalline samples.
- To detect unidentified elements by comparing diffraction data against a database maintained by the International Centre for Diffraction Data (ICPD).
- To determine relative abundance of crystalline composites and provide structural information on unidentified materials.
- To determine strains in crystalline materials.
- To analyse and determine the spreading of crystalline orientations within a crystalline sample.
- To study particles in polycrystalline solids.
- To measure sample purity, etc.

Hence, X-ray diffraction technique has several applications in Material Science, Medical Science, Chemistry, Geology, Environmental Science, Forensic Science and Pharmaceutical Industry for characterizing materials.

3.4 MORPHOLOGICAL STUDY

3.4.1 Scanning Electron Microscopy (SEM) Technique

Scanning electron microscope is a tool to create magnified images by using electrons instead of light waves. SEM produces high resolution images of prepared sample surface by scanning with a focused beam of electrons and X-rays instead of light. Developed scanning electron microscope is used to study and to examine variety of samples in the area of Metallurgy, Geology, Biology, Medicine, Medical and Physical Science communities.

Principle: A thermionically emitted beam of electrons from tungsten filament or a field emission gun is generated and accelerated through a high voltage (20 kV). Such accelerated beam of electrons passes through a system of apertures and electromagnetic lenses to produce a thin beam of electrons. Then a thin beam of electrons scans the surface of the prepared sample or specimen surface using scan coils like the spot in a cathode-ray tube. Pair of scanning coils bends the beam linearly over a rectangular area of the sample surface. The primary electrons interface the surface with inelastic scattered by atoms in the sample. This scattering effectively spreads primary beam and fills tear-drop shaped volume extended about 1 to 5 μm into the surface. This interaction leads to emission of electrons and X-rays and SEM equipped detector produces an image. The resolution of SEM depends on the size of the electrons spot, magnetic electron optical system and interaction volume that creates the scanning beam. The SEM analysis on nanomaterials samples can be performed in order to investigate the morphology and distribution of nanoparticles. A schematic diagram of SEM is shown in Fig. 3.3.

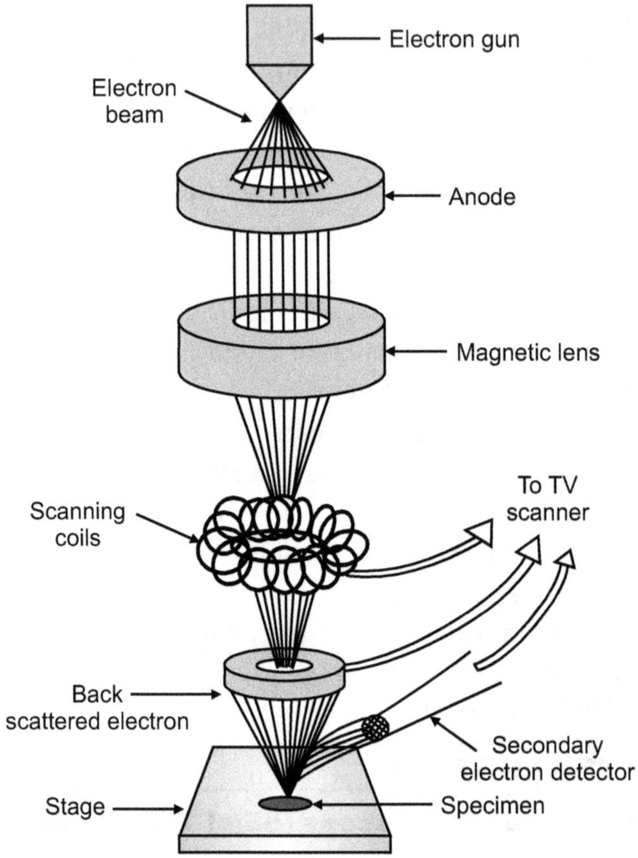

Fig. 3.3: Schematic diagram of an SEM technique

Advantages of Scanning Electron Microscopy:

- Scanning electron microscope instrument works very fast to generate digital data.
- The 3-D, topographical imaging and versatile information are obtained from SEM using different detectors.
- It is easy to operate with the proper training and advances in computer technology.
- Its associated software makes operation user-friendly.

Disadvantages of Scanning Electron Microscopy:

- Scanning electron microscope is expensive.
- Scanning electron microscope requires maintenance that involves keeping constant current and voltage to electromagnetic coils.
- Scanning electron microscope has risk of radiation exposure related with scatter electrons from surface of sample.
- Scanning electron microscope has limited use in case of inorganic and solid samples.

Applications of Scanning Electron Microscopy:

- It is used to study the surface morphology of nanomaterials.
- It is used for semiconductor inspection in industries.
- It is used for inspection of microchips for computers in technology.

- It is used to analyze elemental composition of nanomaterials.
- It is used to detect and analyze surface fractures, to provide information in microstructures, to examine surface contaminations, etc.
- It is used in quantitative chemical analyses and in identification of crystalline structures.

3.4.2 Transmission Electron Microscopy (TEM) Technique

TEM is a microscopic technique having many applications in Cancer Research, Virology, Materials Sciences, Pollution, Semiconductor Research, etc. of both physical and biological scientific fields. This technique detects the internal structure of solids by providing details of morphological fine structures.

Principle: Transmission electron microscope (TEM) is a technique operated under vacuum pressure in the order of 10^{-7} Torr to produce electrons from electron gun and accelerating electrons through a high 60 to 300 kV voltage. Such beam of electrons is partly transmitted through and interacts with an ultra-thin sample. Simultaneously part of electron beam is scattered either inelastically or elastically. These elastically scattered electrons undergo diffractions from the atomic planes of crystal and inelastic electrons arise from the phonon, and X-rays. This transmitted and interacting beam contains information about electron phase, density, periodicity, general background intensity, etc. An image is designed due to the interaction of the electrons transmitted through the specimen. Such magnified image is focused onto sensor or fluorescent screen or layer of photographic film or CCD camera for detection as shown in Fig. 3.4.

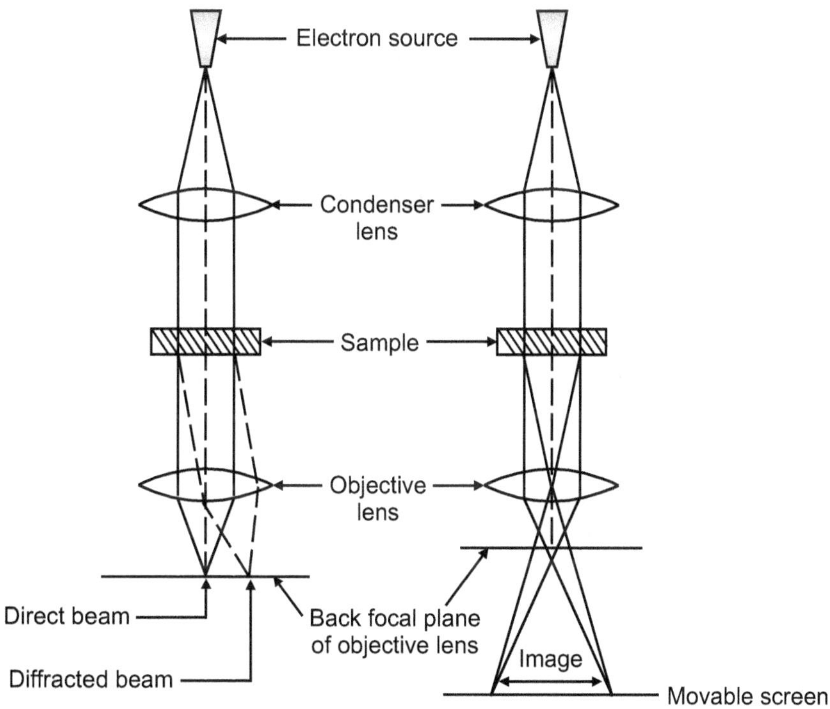

Fig. 3.4: Schematic representation of TEM

Advantages of Transmission Electron Microscope:

- The sample may not be damaged by the electron beam in transmission electron microscopy techniques.
- Transmission electron microscopy technique provides details of morphological fine structures.
- Transmission electron microscopy technique detects internal structure of solids.

Disadvantages of Transmission Electron Microscope:

There are a number of drawbacks of the transmission electron microscope technique.

- The structure of the sample may be changed during the preparation of process in transmission electron microscope analysis.
- The region analyzed may not be characteristic of whole sample in TEM analysis.
- It is a time consuming process.
- In transmission electron microscopy analysis, field of view is relatively small.

EXERCISE

Short-Answer Questions:

1. State the principle of UV-visible spectroscopy.
2. State the principle of X-ray diffraction technique.
3. State the principle of scanning electron microscopy.
4. State the principle of transmission electron microscopy.
5. State Beer-Lambert's law for absorptance, transmittance and reflectance.
6. Give any two advantages of X-ray diffraction technique.
7. Give any two advantages of scanning electron microscopy.
8. Give any two advantages of transmission electron microscopy.
9. State Scherer formula to determine particle size.
10. Give any two disadvantages of X-ray diffraction technique.
11. Give any two disadvantages of scanning electron microscopy.
12. Give any two disadvantages of transmission electron microscopy.
13. Draw schematic diagram for SEM.
14. Give any two applications of UV-visible spectroscopy.
15. Give any two applications of X-ray diffraction technique.
16. Give any two applications of scanning electron microscopy.
17. Give any two applications of transmission electron microscopy.
18. Draw schematic diagram for SEM.

Long-Answer Questions:

1. Describe UV-Visible spectroscopy with neat diagram.
2. Describe X-ray diffraction technique in brief. Also give its advantages and disadvantages.
3. Describe scanning electron microscopy with neat diagram. Also give its applications.
4. Describe transmission electron microscopy with neat diagram. Also give its applications.

☆☆☆

Chapter 4 ...

PROPERTIES OF NANOMATERIALS

- Introduction
- Mechanical Properties of Nanomaterials
- Electrical Properties of Nanomaterials
- Magnetic Properties of Nanomaterials
- Thermal Properties of Nanomaterials
- Optical Properties of Nanomaterials
- Solubility and Melting point of Nanomaterials

4.1 INTRODUCTION

The small size and shape of the material and interactions between nanomaterials decide the properties of the material. Most of the materials have critical size less than 100 nm. Due to nanometer size, nanomaterials have many novel applications. Many novel applications of the nanomaterials rise from novel properties. Nanomaterials are depending on size dependent properties and they can be recognized as follows;

1. Chemical properties - Reactivity, Catalysis
2. Thermal properties - Melting temperature
3. Mechanical properties - Adhesion, Capillary forces
4. Optical properties - Absorption and Scattering of light
5. Electrical properties - Tunneling of current
6. Magnetic properties - Super paramagnetic effect

4.2 MECHANICAL PROPERTIES OF NANOMATERIALS

"Mechanical Properties of Nanoparticles" deal with bulk metallic and ceramic materials, influence of porosity, influence of grain size, super-plasticity, filled polymer composites, particle filled polymers, polymer-based nanocomposites filled with platelets, carbon nanotube-based composites. The discussion of mechanical properties of nanomaterials is, in to some extent, only of quite basic interest, the reason being that it is problematic to produce macroscopic bodies with a high density and a grain size in the range of less than 100 nm. However, two materials neither of which is produced by pressing and sintering have attracted much greater interest as they will undoubtedly achieve industrial importance.

These materials are polymers which contain nanoparticles or nanotubes to improve their mechanical behaviour, and severely plastic-deformed metals, which exhibit astonishing properties. However, due to their larger grain size, the later are generally not accepted as nanomaterials. Experimental studies on the mechanical properties of bulk nanomaterials are generally impaired by major experimental problems in producing specimens with exactly defined grain sizes and porosities. Therefore, model calculations and molecular dynamic studies have major impact for understanding the mechanical properties of these materials.

Filling polymers with nanoparticles or nanorods and nanotubes, respectively leads to significant improvements in their mechanical properties. Such improvements depend heavily on the type of the filler and the way in which the filling is conducted. The latter point is of special importance, as any specific advantages of a nanoparticle at filler may be lost if the filler forms aggregates, thereby mimicking the large particles. Particulate filled polymer-based nano-composites exhibit a broad range of failure strengths and strains. This depends on the shape of the filler, particles or platelets and on the degree of agglomeration. In this class of material, polymers filled with silicate platelets exhibit the best mechanical properties and are of the greatest economic relevance. The larger the particles of the filler or agglomerates, the poorer are the properties obtained. Although, potentially, the best composites are those filled with nanofibers or nanotubes, experience teaches that sometimes such composites have the least ductility. On the other hand, by using carbon nanotubes it is possible to produce composite fibers with extremely high strength and strain at rupture. Among the most exciting nano-composites are the polymer ceramic nanocomposites, where the ceramic phase is platelet shaped. This type of composite is preferred in nature, and is found in the structure of bones, where it consists of crystallized mineral platelets of a few nanometers thickness that are bound together with collagen as the matrix. Composites consisting of a polymer matrix and defoliated phyllosilicates exhibit excellent mechanical and thermal properties.

4.3 ELECTRICAL PROPERTIES OF NANOMATERIALS

Electrical Properties of Nanoparticles" discuss about fundamentals of electrical conductivity in nanotubes and nanorods, carbon nanotubes, photoconductivity of nanorods, electrical conductivity of nanocomposites. The interesting method which can be used to demonstrate the steps in conductance is the mechanical thinning of a nanowire and measurement of the electrical current at a constant applied voltage. The important point is that, with decreasing diameter of the wire, the number of electron wave modes contributing to the electrical conductivity is becoming increasingly smaller by well-defined quantized steps. In electrically conducting carbon nanotubes, only one electron wave mode is observed which transport the electrical current. As the lengths and orientations of the carbon nanotubes are different, they touch the surface of the mercury at different times, which provides two sets of information: (i) the influence of carbon nanotube length on the resistance; and (ii) the resistances of the different nanotubes. As the nanotubes have different lengths, then with increasing protrusion of the fiber bundle an increasing number of carbon nanotubes will touch the surface of the mercury droplet and contribute to the electrical current transport.

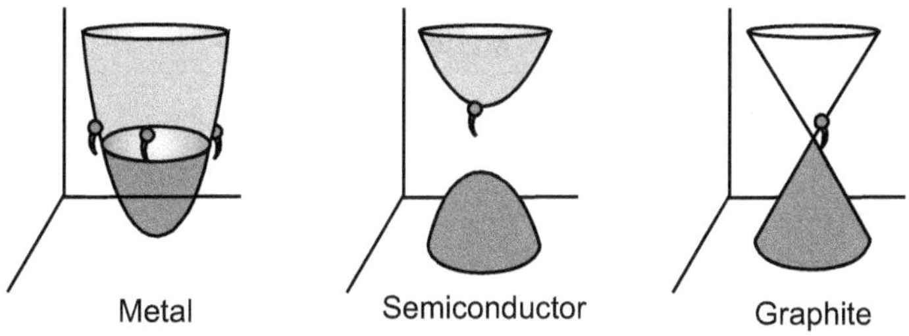

Fig. 4.1: Electrical behaviour of nanotubes

4.4 MAGNETIC PROPERTIES OF NANOMATERIALS

Bulk gold (Au) and Platinum (Pt) are non-magnetic, but at the nano size are magnetic. Surface atoms are not only different to bulk atoms, but also be modified by interaction with other chemical species, that is, by capping the nanoparticles. This phenomenon opens the possibility to modify the physical properties of the nanoparticles by capping them with appropriate molecules.

Actually, it should be possible that non-ferromagnetic bulk materials exhibit ferromagnetic like behaviour when prepared in nano range. One can obtain magnetic nanoparticles of Pd, Pt and the surprising case of Au (that is diamagnetic in bulk) from non-magnetic bulk materials. In the case of Pt and Pd, the ferromagnetism arises from the structural changes associated with size effects. However, gold nanoparticles become ferromagnetic when they are capped with appropriate molecules: the charge localized at the particle surface gives rise to ferromagnetic behaviour. Surface and the core of Au nanoparticles with 2 nm in diameter show ferromagnetic and paramagnetic character, respectively. The large spin-orbit coupling of these noble metals can yield to a large anisotropy and therefore exhibit high ordering temperatures. More surprisingly, permanent magnetism was observed up to room temperature for thiol-capped Au nanoparticles. For nanoparticles with sizes below 2 nm, the localized carriers are in the 5d band. Bulk Au has an extremely low density of states and becomes diamagnetic, as is also the case for bare Au nanoparticles. This observation suggested that modification of the band structure by chemical bonding can induce ferromagnetic like character in metallic clusters.

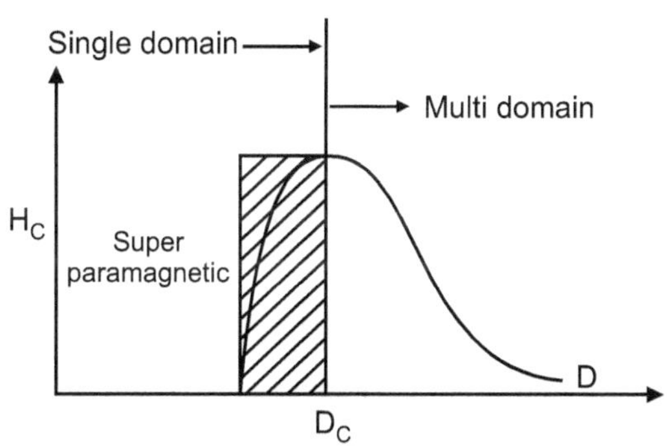

Fig. 4.2: Magnetic properties of nanostructured materials

4.5 THERMAL PROPERTIES OF NANOMATERIALS

Engineering materials are important in everyday life because of their versatile structural properties. Other than these properties, they do play an important role because of their physical properties. Prime physical properties of materials include: electrical properties; thermal properties; magnetic properties; and optical properties. The thermal properties of engineering materials are diverse, and so are their uses in different applications. Thermal properties are: Heat capacity, Thermal expansion, Thermal conductivity and Thermal stresses.

1. Heat Capacity:

Heat capacity is a measure of the ability of the material to absorb thermal energy. Potential energy of a solid material is stored as its heat energy. Temperature of a solid is a measure its potential energy. External energy required to increase temperature of a solid mass is known as the heat capacity of the material. The heat capacity (C) of a system is the ratio of the heat added to the system, or withdrawn from the system, to the resultant change in the temperature. It is an ability to absorb heat energy.

$$C = \frac{dQ}{dt}$$

Heat capacity has units as J/mol-K or cal/mol-K. Heat capacity is not an intrinsic property i.e. it changes with material volume/mass.

At low temperatures, vibrational heat contribution of heat capacity varies with temperature as follows:

$$C_v = AT^3$$

The above relation is not valid above a specific temperature known as Debye temperature. The saturation value is approximately equal to 3R.

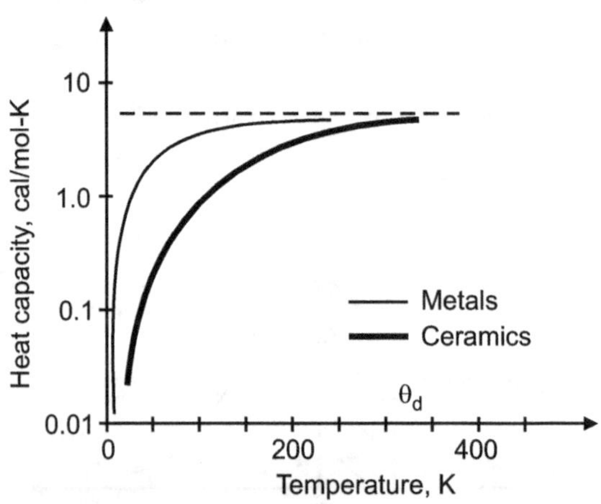

Fig. 4.3 (a): Heat capacity of nanostructured materials

2. Thermal Expansion:

Materials expand when heated and contract when cooled. Rising temperature results in the increase of the average amplitude of atomic vibrations. For a harmonic potential, this corresponds to the increase in the average value of inter-atomic separation called thermal expansion. Linear coefficient of thermal expansion (α) is defined as the change in the dimensions of the material per unit length.

$$\alpha = \frac{l_f - l_0}{l_0 (T_f - T_0)} = \frac{\Delta l}{l_0 \Delta T} = \frac{\varepsilon}{\Delta T}$$

α has unit as $^{\circ}C^{-1}$.

Values of α are as follows:

For metals: $5 - 25 \times 10^{-6}$

For ceramics: $0.5 - 15 \times 10^{-6}$

For polymers: $50 - 400 \times 10^{-6}$

Changes in dimensions with temperature are due to change in inter-atomic distance, rather than increase in vibrational amplitude.

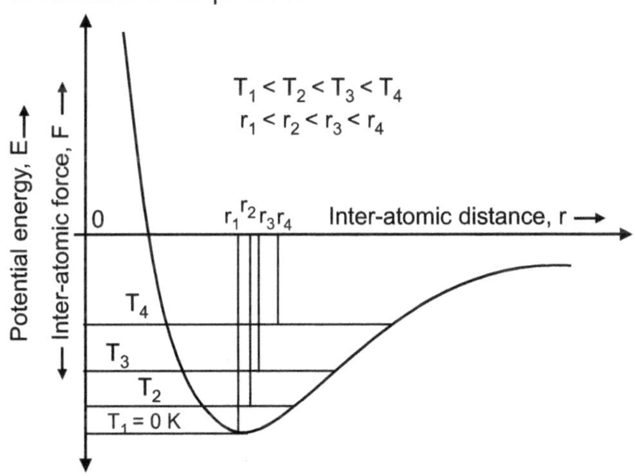

Fig. 4.3 (b): Interatomic force of nanostructured materials

3. Thermal Conductivity:

Thermal conductivity is the ability of a material to transport heat energy through it from high temperature region to low temperature region. Heat energy transported through a body with thermal conductivity 'K' is

$$Q = KA \frac{\Delta T}{\Delta l}$$

It is a microstructure sensitive property. It has units as W/m-K.

Its value ranges as follows;

For metals: 20 - 400

For ceramics: 2 - 50

For polymers: of the order of 0.3

4. Thermal Stresses:

Stresses due to change in temperature or due to temperature gradient are termed as thermal stresses. Thermal stresses in a constrained body will be of compressive nature if it is heated, and vice versa.

$$\sigma_{thermal} = \alpha E \Delta T$$

4.6 OPTICAL PROPERTIES OF NANOMATERIALS

One of the most fascinating and useful aspects of nanomaterials is their optical properties. Applications based on optical properties of nanomaterials include optical detector, laser, sensor, imaging, phosphor, display, solar cell, photo-catalysis, photo electrochemistry and biomedicine. The optical properties of nanomaterials depend on parameters such as feature size, shape, surface characteristics, and other variables including doping and interaction with the surrounding environment or other nanostructures. Likewise, shape can have dramatic influence on optical properties of metal nanostructures. Fig. 4.4 exemplifies the difference in the optical properties of metal and semiconductor nanoparticles. With the CdSe semiconductor nanoparticles, a simple change in size alters the optical properties of the nanoparticles. When metal nanoparticles are enlarged, their optical properties change only slightly as observed for the different samples of gold nanospheres. However, when anisotropy is added to the nanoparticle, such as growth of nanorods, the optical properties of the nanoparticles are changed dramatically.

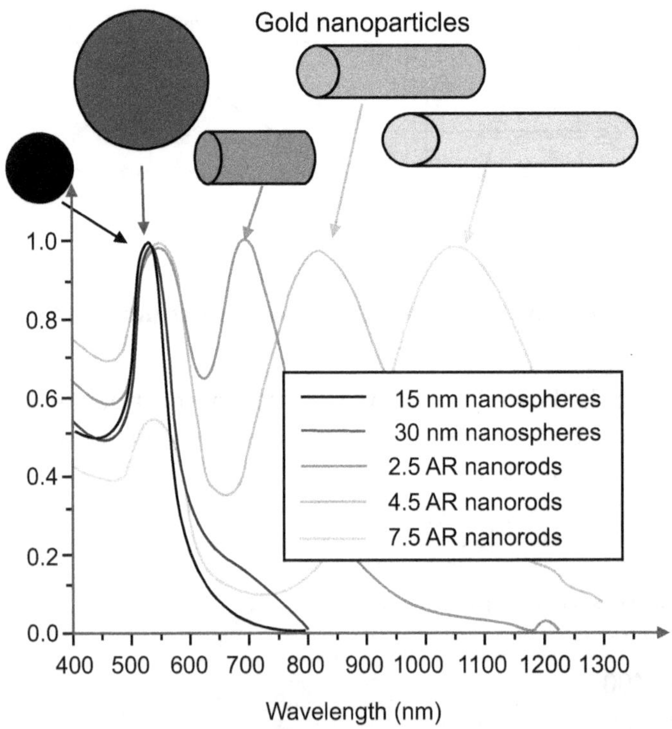

Fig. 4.4: Absorption spectra of various sizes and shapes of gold nanoparticles

4.7 SOLUBILITY AND MELTING POINT OF NANOMATERIALS

Solubility of Nanomaterials:

Nanoparticles with a high degree of uniformity will be produced using modern synthetic method. When the nanoparticles are surface ligated with organic compounds, they are stable against irreversible aggregation. Nanoparticle colloidal suspensions can act as solutions with reversible, temperature dependent solubility. So, a suspension of nanoparticle monomers can be in equilibrium with precipitated aggregates of nanoparticles. Like a solution, the nanoparticle suspension has a phase diagram with one- and two-phase regions. In the two phase region, the supernatant is the mixture of nanoparticle monomers in the solvent. Aggregates of nanoparticles settle out under gravity to form the precipitate.

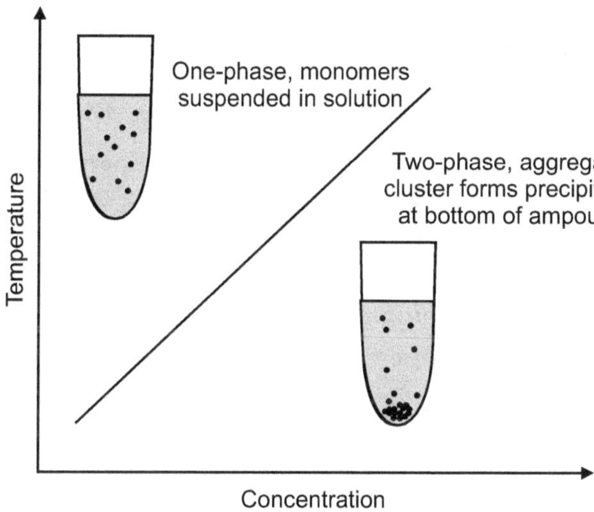

Fig. 4.5: Schematic phase diagram of nanoparticles in solution

Melting Point of Nanomaterials:

Melting point of nanomaterials provides the basis of size and shape dependent melting temperature change in the metal nanoparticle. Many physical properties of materials like melting point change when the physical size of the material approaches the micro and nanoscales. Melting-point depression is a term referring to the phenomenon of reduction of the melting point of a material with reduction of its size. This phenomenon is prominent in nanoscale materials which melt at temperatures hundreds of degrees lower than bulk materials. However as the dimensions of a material decrease towards the atomic scale, the melting temperature scales with the material dimensions. Melting point depression is most evident in nanowires, nanotubes and nanoparticles, which all melt at lower temperatures than bulk amounts of the same material. Changes in melting point occur because nanoscale materials have a much larger surface-to-volume ratio than bulk materials, drastically altering their thermodynamic and thermal properties. As the metal particle size decreases, the melting temperature also decreases.

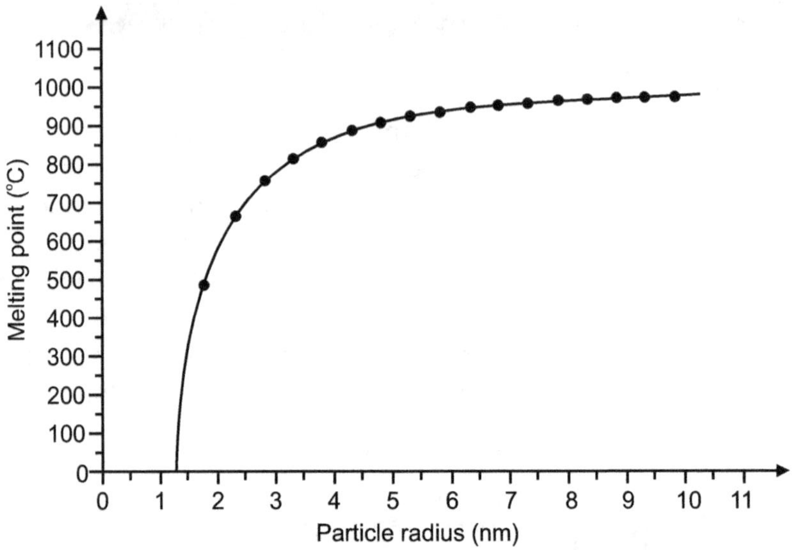

Fig. 4.6: Schematic melting point of nanoparticles

EXERCISE

Short-Answer Questions:

1. How nanomaterials chemical properties are recognized?
2. How nanomaterials optical properties are recognized?
3. How nanomaterials magnetic properties are recognized?
4. How nanomaterials mechanical properties are recognized?
5. How nanomaterials thermal properties are recognized?
6. What do you meant by thermal stress?
7. What do you meant by thermal conductivity?
8. What do you meant by thermal expansion?
9. What do you meant by heat capacity?

Long-Answer Questions:

1. Explain optical properties of nanosize materials with examples.
2. Explain magnetic properties of nanosize materials with examples.
3. Explain electrical properties of nanosize materials with examples.
4. Explain mechanical properties of nanosize materials with examples.
5. Explain thermal properties of nanosize materials with examples.
6. Explain any two properties of nanosize materials with examples.
7. Write a short note on solubility and melting point of nanomaterials.

☆☆☆

Chapter 5 ...

SPECIAL NANOMATERIALS

- Introduction
- Special Nanomaterials
- Carbon Nanotubes
- Quantum Dots
- Nanocrystalline ZnO and TiO_2

5.1 INTRODUCTION

Novel discoveries, laws, phenomena, technical decisions, solutions, and inventions will be possibly made only in specially designed and assembled artificial nanostructures to be fabricated by future material science. Material science concept is shown in Fig. 5.1 illustrating the inherent interconnection between the composition, structure, properties, technology and applications.

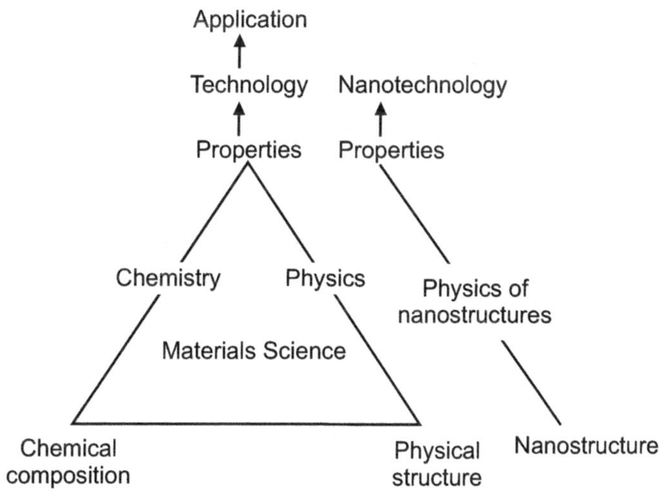

Fig. 5.1: Fundamental triad of Materials Science

Nanoscience deals with synthesis, characterization, exploration, and exploitation of nanostructured materials. Nanophysics means physics of nanostructures which is relatively new areas of science arisen in last decade of past century after discovery of fullerenes and nanotubes. It is introduction into more extent interdisciplinary integrated modern science known now as nanotechnology rapidly developing. The peculiar properties of nanostructures are emphasized to be result of size effects, external and internal, classical and quantum dots. Numerous applications are considered including micro-lasers, photonic crystals, probe microscopy, left-handed materials with negative refractive index, etc.

5.2 SPECIAL NANOMATERIALS

Atoms, molecules, clusters, fullerenes, supra-molecular structures, their crystals, nanotubes, nanowires, nanorods, their arrays and photonic crystals serve as nanotechnology objectives. Fullerenes and atomic clusters are the smallest zero-dimensional (0D) nanostructures called quantum dots possessing the properties inherent for nanomaterial rather than for single atom. For fullerenes it should mean not only the buckyball C_{60}, but the multitude of another carbon C_n and non-carbon clusters and metcarbes Me@C_n. Presently a number of experimental nanodevices are developed on this base, e.g. the switches, diodes, transistors, amplifiers, sensors, optical filters, solar cells, magneto-optical recorders, etc. Nanotubes, nanorods, nanowires, nanofibers manifest more advanced and promising properties as being the 1D quantum wires nanoscopic in diameter but microscopic in length. Their unique properties stem from capability of the ring and cylindrical types of acoustic and electromagnetic waves to propagate that makes them a unique nano-laboratory for research of quantum resonance phenomena. All stated above also concern to noncarbon 1D nanowires and nanotubes based on boron-nitride, oxides, chalcogenides, dichalcogenides, and some other III–V and II–VI compounds possessing the most manifold physical-chemical characteristics.

5.3 CARBON NANOTUBES

The discovery of carbon nanotubes (CNT) in 1991 opened up a new era in materials science. Nanotubes (NT) are recognized as one of the base for the future nano-electronics. A carbon nanotube can be as thin as a few nanometers. A carbon nanotube is a tube-shaped material, made of carbon, having a diameter measuring on the nanometer scale. Nanotubes are the 1D quantum cylinders nanoscopic in diameter while micro/macroscopic in length so they are quantum macro-crystals possessing potential unique properties. CNT are unique because the bonding between the atoms is very strong and the tubes can have extreme aspect ratios. These incredible structures have an array of fascinating electronic, magnetic and mechanical properties. CNT are at least 100 times stronger than steel but only $1/6^{th}$ as heavy. Nanotubes can conduct heat and electricity far better than copper. CNT are used in polymers to control conductivity and added to anti-static packaging.

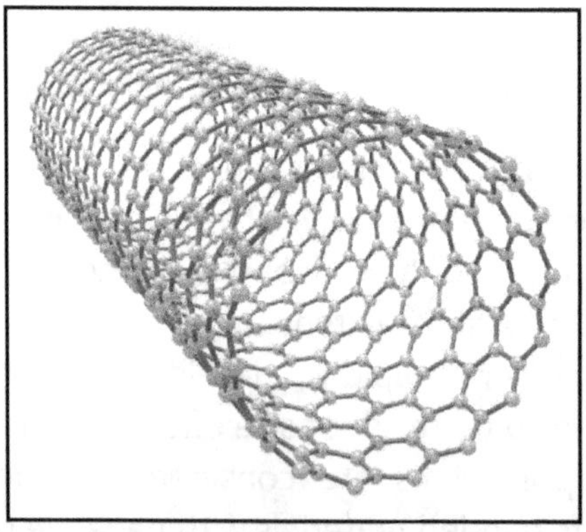

Fig. 5.2: Carbon nanotube in rolling form

Carbon nanotubes have many structures, differing in length, thickness, and number of layers. The characteristics of nanotubes can be different depending on how the graphene sheet has rolled up to form the tube causing it to act either metallic or as a semiconductor. There are many different types of carbon nanotubes, but they are normally categorized as either single-walled (SWNT) or multi-walled nanotubes (MWNT). A single-walled carbon nanotube is just like a regular straw. It has only one layer, or wall. Multi-walled carbon nanotubes are a collection of nested tubes of continuously increasing diameters. They can range from one outer and one inner tube (a double-walled nanotube) to as many as 100 tubes (walls) or more. Each tube is held at a certain distance from either of its neighboring tubes by interatomic forces. Note that, growth of carbon nanotubes in general always requires the presence of a catalyst.

Properties of Carbon Nanotubes:

1. CNTs have high electrical conductivity.
2. CNTs have very high tensile strength.
3. CNTs are highly flexible - can be bent considerably without damage.
4. CNTs are very elastic, ~18% elongation to failure.
5. CNTs have high thermal conductivity.
6. CNTs have a low thermal expansion coefficient.
7. CNTs are good electron field emitters.

Applications of Carbon Nanotubes:

1. **Thermal Conductivity:** CNTs provide opportunity to modify their structure and to optimize their solubility and dispersion.
2. **Field Emission:** CNTs receiving considerable interest is in field-emission flat-panel displays.
3. **Conductive Properties:** CNTs have EMI/RFI shielding composites; coatings f enclosures, gaskets, and other uses electrostatic dissipation (ESD) and antistatic materials and (even transparent!) conductive coatings and radar-absorbing materials for low-observable ("stealth") applications.
4. **Energy Storage:** CNTs have applications in a variety of fuel cell components. CNTs have the highest reversible capacity of any carbon material for use in lithium-ion batteries.
5. **Conductive Adhesive:** Adhesives, potting compounds, and coaxial cables and other types of connectors are the use of CNTs.
6. **Molecular Electronics:** CNTs are demonstrated as switches themselves.
7. **Thermal Materials:** Heat transformation application is found in electronics particularly advanced computing, where uncooled chips now routinely reach over 100°C.
8. **Structural Applications:** Stiffness, toughness, and strength properties lead to a wealth of applications exploiting them, including advanced composites requiring high values of one or more of these properties.
9. **Fibers and Fabrics:** CNTs are being used to make textiles stain resistant.

10. **Biomedical Applications:** Single strand of DNA can be bonded to a nanotube, which can then be successfully inserted into a cell.
11. **Air and Water Filtration:** CNTs filters can not only block the smallest particles, but also kill most bacteria.

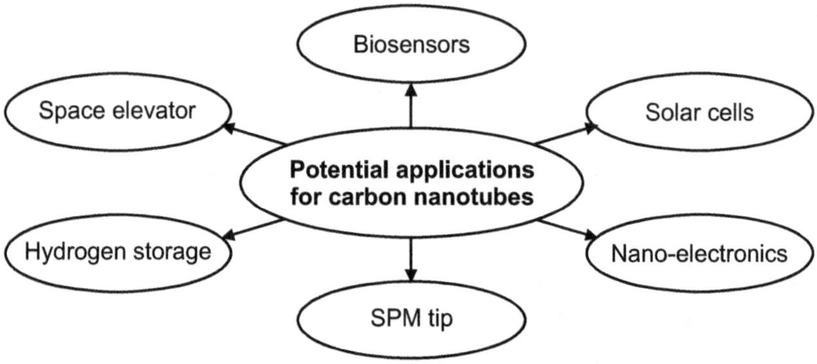

Fig. 5.3: Applications of CNTs (Carbon Nano-Tubes)

5.4 QUANTUM DOTS

Quantum dots (QD) are tiny man-made crystals. They are so small that you can't see them with a typical microscope. In fact, they are 10,000 times narrower than a human hair. That's incredibly small but don't let their size fool you. Quantum dots (QD) are very small semiconductor particles, only several nanometers in size, so small that their optical and electronic properties differ from those of larger particles. Quantum dots obey quantum mechanical principle of quantum confinement. Quantum dots exhibit energy band gap that determines required wavelength of radiation absorption and emission spectra. Energy band gap increases with decreasing size of quantum dots. They are a central theme in nanotechnology. Many types of quantum dots will emit light of specific frequencies if electricity or light is applied to them, and these frequencies can be precisely tuned by changing the dots' size, shape and material, giving rise to many applications.

Quantum dots are actually very powerful devices and their size gives them a unique ability: to convert light into nearly any colour in the visible spectrum with very high efficiency. Quantum dots' ability to precisely convert and tune a spectrum of light makes them ideal for LCD displays. From smart phones to tablets to TVs, we can make all the colours we see even better by remixing white light into red, green and blue components. Until now, the white light that LCDs have had to work with wasn't very good, it contained a lot of blue and yellow but not very much red or green. This meant displays had to waste a lot of energy to make enough red and green for a bright display while also making for broad primary colours. With quantum dots we can design an ideal spectrum of white light for an LCD, one that contains only the red, green and blue that the display needs to make a great image. The precise spectrum created by the dots makes colours pure. And since we are only making the colours the display needs, we can use less power. The result is a display that's brighter, more power efficient and incredibly vibrant.

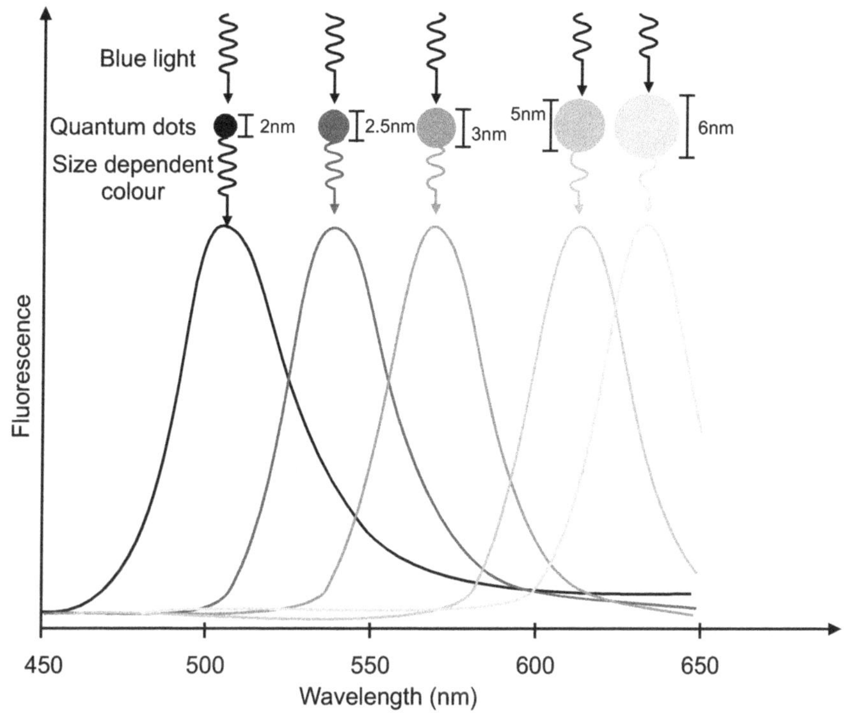

Fig. 5.4: Quantum dots and their colours

Applications of Quantum Dots:

Quantum dots have found use in the following applications:

1. To produce TVs with much higher peak brightness.
2. Quantum dots improve colour accuracy. The light produced by quantum dots is so closely tied to their size that they can be tuned very precisely to emit the exact kind of light needed. This means purer, cleaner whites and more precise colours.
3. Useful for monitoring cancerous cells and providing a means to better understand its evolution.
4. Used to produce inexpensive, industrial quality white light.
5. Integrate quantum dots into dust that tracks enemies and protection against friendly-fire events.
6. Inject QD into liquid mixtures, fabrics, polymer matrices, etc. to increase the ability to specifically control absorption and emission spectra to produce unique validation signatures.
7. Improved battery life in mobile gadgets.
8. Achieving attractive colours in stained glasses is the oldest application.
9. Composite Applications of quantum dots.
10. Production of laser diodes with low temperature sensitivity and/or very low threshold pump power.

11. In white light-emitting diodes (LEDs): quantum dots emit green and red light when excited with a blue or near-ultraviolet LED, thus achieving overall a white colour tone.
12. In semiconductor saturable absorber mirrors (SESAMs).
13. As absorbers with very low saturation fluence.
14. It is possible to contain quantum dot absorbers in a glass matrix.
15. As single-photon emitters in the perspective of quantum nano-photonics and quantum cryptography.
16. As components of solar cells (Gratzel cells) and highly sensitive photodetectors.
17. As fluorescent biological labels in the biological field.

5.5 NANOCRYSTALLINE ZnO AND TiO_2

A nanocrystalline (NC) material is a polycrystalline material with a crystallite size of only a few nanometers. These materials fill the gap between amorphous materials without any long range order and conventional coarse-grained materials. Nanocrystal is a material particle having at least one dimension smaller than 100 nanometers and composed of atoms in either a single-crystalline or poly-crystalline arrangement. Nanocrystals made with zeolite are used to filter crude oil onto diesel fuel.

5.5.1 Nanocrystalline ZnO

Zinc oxide is an inorganic compound with the formula ZnO. It usually appears as a white powder, nearly insoluble in water. The powder is widely used as an additive into numerous materials and products including plastics, ceramics, glass, cement, rubber (e.g. car tyres), lubricants, paints, ointments, adhesives, sealants, pigments, foods (source of Zn nutrient), batteries, ferrites, fire retardants, etc. The ZnO is present in the Earth crust as a mineral zincite. The most ZnO used commercially is produced synthetically. Zinc oxide crystallizes in three forms: hexagonal wurtzite, cubic zincblende, and the rarely observed cubic rocksalt. The wurtzite structure is most stable and thus most common at ambient conditions. The zinc blende form can be stabilized by growing ZnO on substrates with cubic lattice structure. In both cases, the zinc and oxide are tetrahedral. The rocksalt NaCl-type structure is only observed at relatively high pressures (10 GPa).

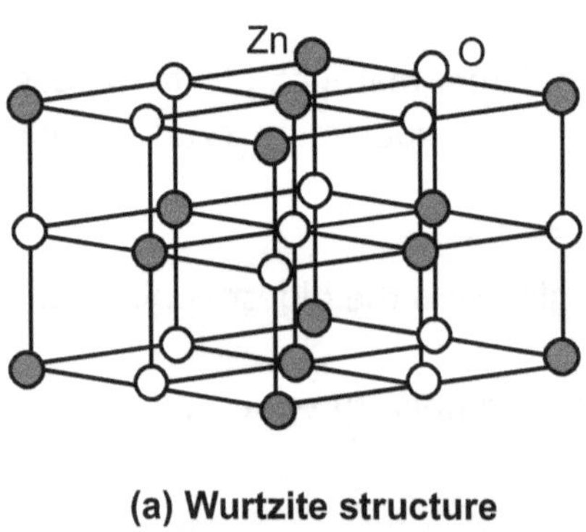

(a) Wurtzite structure

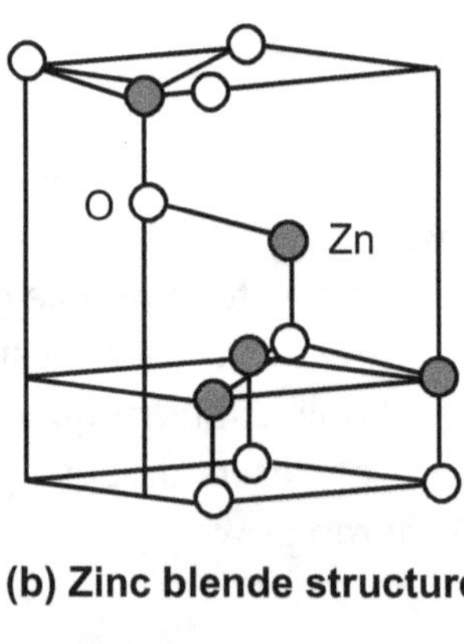

(b) Zinc blende structure

Fig. 5.5: Crystal structures of ZnO

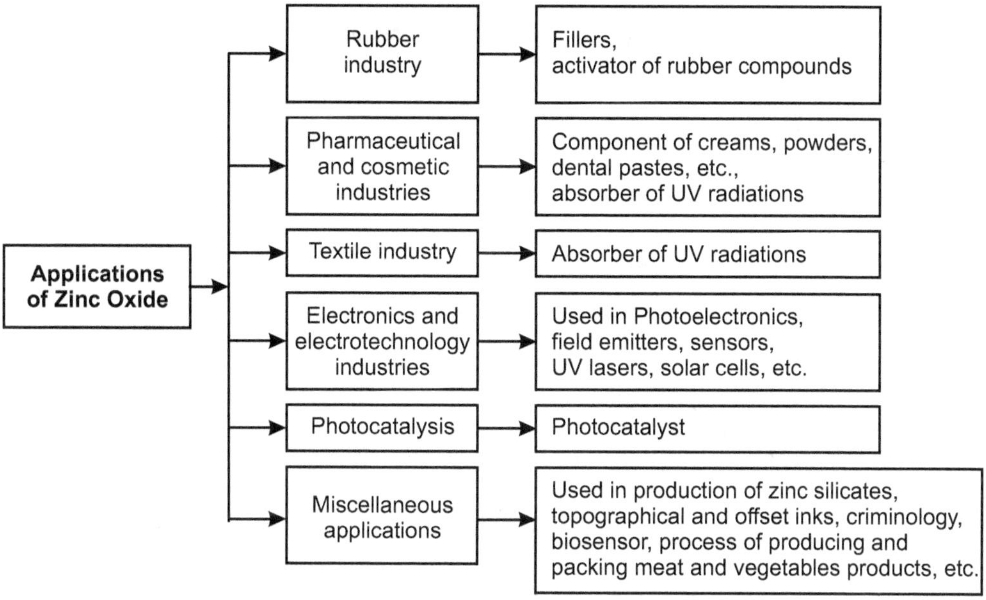

Fig. 5.6

5.5.2 Nanocrystalline TiO_2

Nanocrystalline semiconductor TiO_2 particles are of interest due to their unique properties and several potential technological applications such as photocatalysis, sensors, solar cells and memory devices. TiO_2 exists in three polymorphic phases: rutile (tetragonal density = 4.25 g/cm^3), anatase (tetragonal, 3.894 g/cm^3) and brookite (orthorhombic, 4.12 g/cm^3). Both anatase and rutile have tetragonal crystal structures, but belong to different space groups. Among the three above mentioned crystal structures of TiO anatase owing to its higher photo catalytic activity is commonly used for photo catalysis. This higher photo catalytic activity is related to its lattice structure. Each Ti atom is coordinated to six oxygen atoms in anatase tetragonal unit cell. TiO_2 is stable in aqueous media and is tolerant of both acidic and alkaline solutions. It is cheap, recyclable, reusable and relatively simple to produce. It can also be synthesized in nanostructure forms more readily than many other catalysts. Furthermore, its band gap is appropriate to initiate a variety of organic reactions.

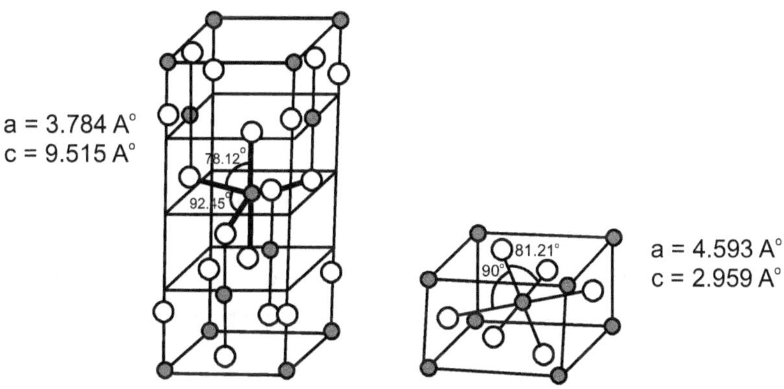

(a) Anatase TiO_2 **(b) Rutile TiO_2**

Fig. 5.7: Crystal structure TiO_2

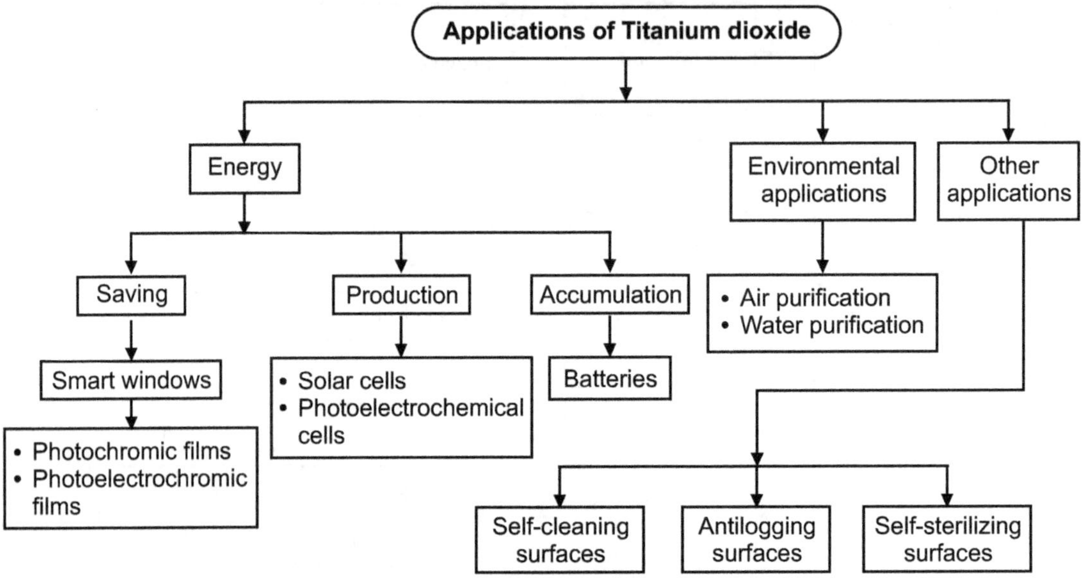

Fig. 5.8

EXERCISE

Short-Answer Questions:

1. What is nanophysics?
2. What is nanoscience?
3. What are carbon nanotubes?
4. Give various properties of carbon nanotubes.
5. State various applications of carbon nanotubes.
6. Define : Quantum dots.
7. State various applications of carbon nanotubes.
8. Define : Nanocrystals.
9. Define : Nanocrystalline materials.

Long-Answer Questions:

1. Write a note on special nanomaterials.
2. What are carbon nanotubes? Explain various properties of it. Also give its applications.
3. Explain quantum dots with neat diagram? Also explain its various applications.
4. Define nanocrystalline materials. Describe nano-crystalline TiO_2 and ZnO.

☆☆☆

Chapter 6 ...

APPLICATIONS OF NANOTECHNOLOGY

- Introduction
- Applications of Nanomaterials
- Applications of Nanotechnology: Nanoelectronics, Medical, Biological, Automobiles Space, Defense, Sports, Cosmetics, Cloth industry

6.1 INTRODUCTION

Nanotechnology is helping to considerably improve, even revolutionize, many technology and industry sectors: information technology, homeland security, medicine, transportation, energy, food safety, and environmental science, and among many others. Many applications of nanotechnology depend on the fact that it is possible to tailor the structures of materials at extremely small scales to achieve specific properties. Using nanotechnology, materials can effectively be made stronger, lighter, more durable, more reactive, more sieve-like, or better electrical conductors, among many other traits. Many everyday commercial products are currently on the market and in daily use that rely on nanoscale materials.

Nanoscale additives to or surface treatments of fabrics can provide lightweight ballistic energy deflection in personal body armor, or can help them resist wrinkling, staining, and bacterial growth.

Clear nanoscale films on eyeglasses, computer and camera displays, windows, and other surfaces can make them water-repellent and residue-repellent, antireflective, self-cleaning, resistant to ultraviolet or infrared light, anti-fog, antimicrobial, scratch-resistant, or electrically conductive.

Nanoscale materials are beginning to enable washable, durable "smart fabrics" equipped with flexible nanoscale sensors and electronics with capabilities for health monitoring, solar energy capture, and energy harvesting through movement.

Light weighting of cars, trucks, airplanes, boats, and space craft could lead to significant fuel savings. Nanoscale additives in polymer composite materials are being used in baseball bats, tennis rackets, bicycles, motorcycle helmets, automobile parts, luggage, and power tool housings, making them lightweight, stiff, durable, and resilient. Carbon nanotube sheets are now being produced for use in next-generation air vehicles. For example, the combination of light weight and conductivity makes them ideal for applications such as electromagnetic shielding and thermal management.

Nano-bioengineering of enzymes is aiming to enable conversion of cellulose from wood chips, corn stalks, unfertilized perennial grasses, etc., into ethanol for fuel.

Cellulosic nanomaterials have demonstrated potential applications in a wide array of industrial sectors, including electronics, construction, packaging, food, energy, health care, automotive, and defense. Cellulosic nanomaterials are projected to be less expensive than many other nanomaterials and, among other characteristics, tout an impressive strength-to-weight ratio.

Nano-engineered materials in automotive products include high-power rechargeable battery systems; thermoelectric materials for temperature control tires with lower rolling resistance; high-efficiency/low-cost sensors and electronics, thin-film smart solar panels and fuel additives for cleaner exhaust and extended range.

Nanostructured ceramic coatings exhibit much greater toughness than conventional wear-resistant coatings for machine parts. Nanotechnology enabled lubricants and engine oils also significantly reduce wear and tear, which can significantly extend the lifetimes of moving parts in everything from power tools to industrial machinery.

Nanoparticles are used increasingly in catalysis to boost chemical reactions. This reduces the quantity of catalytic materials necessary to produce desired results, saving money and reducing pollutants. Two big applications are in petroleum refining and in automotive catalytic converters.

Nano-engineered materials make superior household products such as degreasers and stain removers; environmental sensors, air purifiers, and filters, antibacterial cleansers and specialized paints and sealing products, such a self-cleaning house paints that resist dirt and marks.

10. Nanoscale materials are also being incorporated into a variety of personal care products to improve performance. Nanoscale titanium dioxide and zinc oxide have been used for years in sunscreen to provide protection from the sun while appearing invisible on the skin.

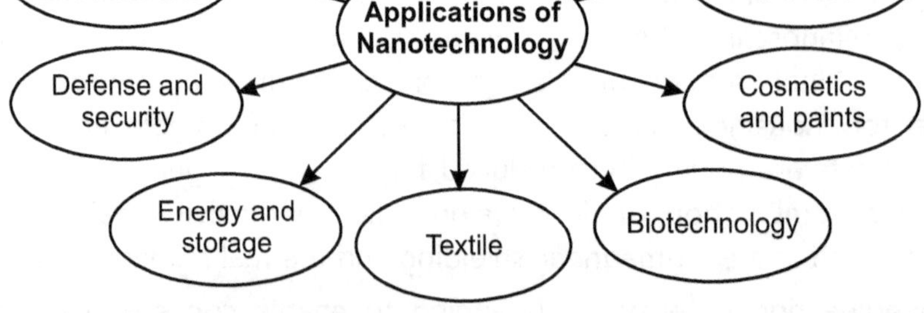

Fig. 6.1: Applications of nanotechnology

6.2 APPLICATIONS OF NANOMATERIALS

Following are the few applications of nanomaterials:

1. **Nanomaterial Applications using Carbon Nanotubes:**

Applications being developed for carbon nanotubes include adding antibodies to nanotubes to form bacteria sensors. A composite nanotube is used to bend the wings of morphing aircraft by applying electric voltage. By adding boron or gold to nanotubes is used to trap oil spills. Coating nanotubes with silicon is used to make anodes that can increase the capacity of Li-ion batteries by up to 10 times.

2. **Nanomaterial Applications using Graphene:**

Applications being developed for graphene include using graphene sheets as electrodes in ultra-capacitors which will have as much storage capacity as batteries but will be able to recharge in minutes, attaching strands of DNA to graphene to form sensors for rapid disease diagnostics, replacing indium in flat screen TVs and making high strength composite materials.

3. **Nanomaterial Applications using Nano-composites:**

Applications being developed for nano-composites include a nanotube-polymer nano-composite to form a scaffold which speeds up replacement of broken bones, making graphene-epoxy nano-composites with very high strength-to-weight ratios, a nano-composite made from cellulous and nanotubes used to make a flexible battery.

4. **Nanomaterial Applications using Nanofibers:**

Applications being developed for nanofibers include stimulating the production of cartilage in damaged joints, piezoelectric nanofibers that can be woven into clothing to produce electricity for cell phones or other devices, carbon nanofibers that can improve the performance flame retardant in furniture.

5. **Nanomaterial Applications using Nanoparticles:**

Applications being developed for nanoparticles include deliver chemotherapy drugs directly to cancer tumors, resetting the immune system to prevent autoimmune diseases, delivering drugs to damaged regions of arteries to fight cardiovascular disease, create photocatalysts that produce hydrogen from water, reduce the cost of producing fuel cells and solar cells, clean up oil spills, water pollution and air pollution.

6. **Nanomaterial Applications using Nanowires:**

Applications being developed for carbon nanotubes include using zinc oxide nanowires in a flexible solar cell, silver chloride nanowires to decompose organic molecules in polluted water, using nanowires made from iron and nickel to make dense computer memory – called Race track memory.

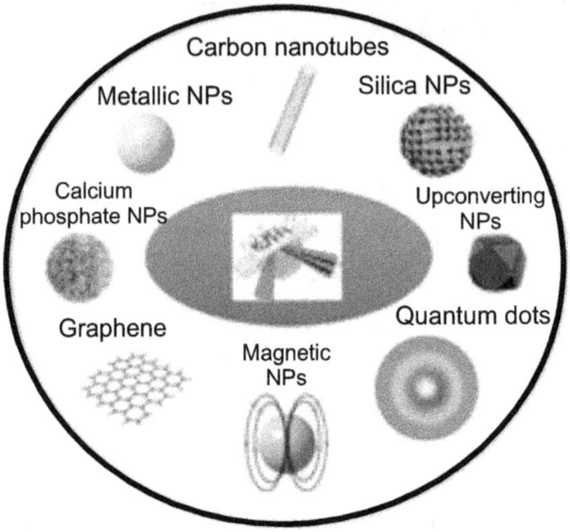

Fig. 6.2: Applications of Nanomaterials

6.3 APPLICATIONS OF NANOTECHNOLOGY

6.3.1 Nanotechnology in Electronics and IT

Nanotechnology has greatly contributed to major advances in computing and electronics, leading to faster, smaller, and more portable systems that can manage and store larger and larger amounts of information. These continuously evolving applications include:

Transistors, the basic switches that enable all modern computing, have gotten smaller and smaller through nanotechnology. At the turn of the century, a typical transistor was 130 to 250 nanometers in size. In 2014, Intel created a 14 nanometer transistor, then IBM created the first seven nanometer transistor in 2015, and then Lawrence Berkeley National Lab demonstrated a one nanometer transistor in 2016. Smaller, faster, and better transistors may mean that soon your computer's entire memory may be stored on a single tiny chip.

Using magnetic random access memory (MRAM), computers will be able to "boot" almost instantly. MRAM is enabled by nanometer-scale magnetic tunnel junctions and can quickly and effectively save data during a system shutdown or enable resume-play features.

Ultra-high definition displays and televisions are now being sold that use quantum dots to produce more vibrant colors while being more energy efficient.

Flexible, bendable, foldable, rollable, and stretchable electronics are reaching into various sectors and are being integrated into a variety of products, including wearable's, medical applications, aerospace applications, and the Internet of Things. Flexible electronics have been developed using, for example, semiconductor nanomembranes for applications in smart phone and e-reader displays. Other nanomaterials like graphene and cellulosic nanomaterials are being used for various types of flexible electronics to enable wearable and "tattoo" sensors, photovoltaics that can be sewn onto clothing, and electronic paper that can be rolled up. Making flat, flexible, lightweight, non-brittle, highly efficient electronics opens the door to countless smart products.

Other computing and electronic products include flash memory chips for smart phones and thumb drives; ultra-responsive hearing aids; antimicrobial/antibacterial coatings on keyboards and cell phone casings; conductive inks for printed electronics for RFID/smart cards/smart packaging; and flexible displays for e-book readers.

Nanoparticle copper suspensions have been developed as a safer, cheaper, and more reliable alternative to lead-based solder and other hazardous materials commonly used to fuse electronics in the assembly process.

6.3.2 Nanotechnology in Medical

Nanotechnology is already broadening the medical tools, knowledge, and therapies currently available to clinicians. Nanomedicine, the application of nanotechnology in medicine, draws on the natural scale of biological phenomena to produce precise solutions for disease prevention, diagnosis, and treatment.

Below are some examples of recent advances in this area:

Commercial applications have adapted gold nanoparticles as probes for the detection of targeted sequences of nucleic acids, and gold nanoparticles are also being clinically investigated as potential treatments for cancer and other diseases.

Better imaging and diagnostic tools enabled by nanotechnology are paving the way for earlier diagnosis, more individualized treatment options, and better therapeutic success rates.

Nanotechnology is being studied for both the diagnosis and treatment of atherosclerosis, or the buildup of plaque in arteries. In one technique, researchers created a nanoparticle that mimics the body's "good" cholesterol, known as HDL (high-density lipoprotein), which helps to shrink plaque.

The design and engineering of advanced solid-state nanopore materials could allow for the development of novel gene sequencing technologies that enable single-molecule detection at low cost and high speed with minimal sample preparation and instrumentation.

Nanotechnology researchers are working on a number of different therapeutics where a nanoparticle can encapsulate or otherwise help to deliver medication directly to cancer cells and minimize the risk of damage to healthy tissue. This has the potential to change the way doctors treat cancer and dramatically reduce the toxic effects of chemotherapy.

Research in the use of nanotechnology for regenerative medicine spans several application areas, including bone and neural tissue engineering. For instance, novel materials can be engineered to mimic the crystal mineral structure of human bone or used as a restorative resin for dental applications. Researchers are looking for ways to grow complex tissues with the goal of one day growing human organs for transplant. Researchers are also studying ways to use graphene nanoribbons to help repair spinal cord injuries; preliminary research shows that neurons grow well on the conductive graphene surface.

Nanomedicine researchers are looking at ways that nanotechnology can improve vaccines, including vaccine delivery without the use of needles. Researchers also are working to create a universal vaccine scaffold for the annual flu vaccine that would cover more strains and require fewer resources to develop each year.

6.3.3 Nanotechnology in Biology

Living organisms are built of cells that are typically 10 μm across. However, the cell parts are much smaller and are in the sub-micron size domain. Even smaller are the proteins with a typical size of just 5 nm, which is comparable with the dimensions of smallest manmade nanoparticles. This simple size comparison gives an idea of using nanoparticles as very small probes that would allow us to spy at the cellular machinery without introducing too much interference. Understanding of biological processes on the nanoscale level is a strong driving force behind development of nanotechnology. Out of plethora of size dependent physical properties available to someone who is interested in the practical side of nanomaterials, optical and magnetic effects are the most used for biological applications. A list of some of the applications of nanomaterials to biology is given below:

- Fluorescent biological labels
- Drug and gene delivery
- Bio detection of pathogens
- Detection of proteins
- Probing of DNA structure
- Tissue engineering
- Tumour destruction via heating (hyperthermia)
- Separation and purification of biological molecules and cells
- MRI contrast enhancement
- Phagokinetic studies

As mentioned above, the fact that nanoparticles exist in the same size domain as proteins makes nanomaterials suitable for bio tagging or labeling. However, size is just one of many characteristics of nanoparticles that it is rarely sufficient if one is to use nanoparticles as biological tags. In order to interact with biological target, a biological or molecular coating or layer acting as a bioinorganic interface should be attached to the nanoparticle. Examples of biological coatings may include antibodies, biopolymers like collagen, or monolayers of small molecules that make the nanoparticles biocompatible. In addition, as optical detection techniques are wide spread in biological research, nanoparticles should either fluoresce or change their optical properties. The approaches used in constructing nano-biomaterials are schematically presented in Fig. 6.3.

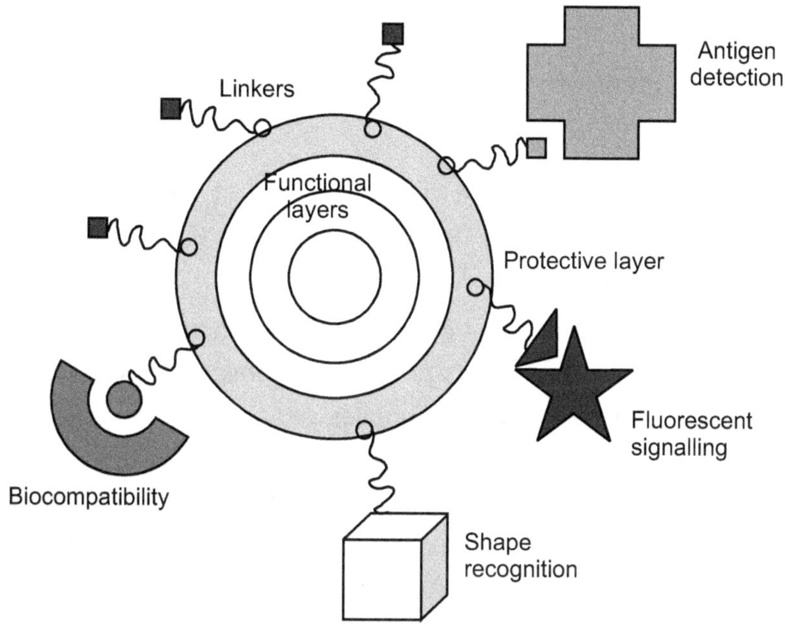

Fig. 6.3: Typical configurations utilized in nano-biomaterials applied to biological problems

6.3.4 Nanotechnology in Automobiles

Nanotechnology had began 30 years ago. It is the manipulation of matter on an atomic and molecular scale. Nanotechnology works with material, devices and other structures with at least one dimension sized from 1 to 100 nm. Nanoscience and Nanotechnology can be used across small things and can be used across all other science fields such as chemistry, physics, material science and engineering. The automotive sector is a major consumer of material technologies and nanotechnologies promise to improve the performance of existing technologies significantly. Applications range from already existing paint quality, fuel cells, batteries, wear-resistant tires, lighter but stronger materials, ultra-thin anti-glare layers for windows and mirrors to the futuristic energy harvesting bodywork, fully self-repairing paint, switchable colors, shape-shifting skin.

The basic trends that nanotechnology enables for the automobile are;

- Lighter but stronger materials for better fuel consumption and increased safety.
- Improved engine efficiency and fuel consumption for gasoline-powered cars such as catalysts, fuel additives, lubricants.
- Reduced environmental impact from hydrogen and fuel cell-powered cars.
- Improved and miniaturized electronic systems.
- Better economies for longer service life, lower component failure rate and smart materials for self-repair.

Following are some of the applications of nanotechnology in automobile industry :

- Lotus coatings,
- In windshields,
- Reduction in weight of vehicles,
- Better batteries,
- Fuel cells,
- Catalytic converters,
- Reduction in fuel consumption,

Though the nanotech products are expensive, they increase the performance of the automobile tremendously. Life of parts of the automobile increases. Maintenance of the automotive can be minimized. In automobiles nanomaterial coating can prevent the corrosive materials from corrosion. They decrease the frictional, tear and wear losses and thus they decrease the rate of fuel consumption. They are extremely small and light weighted, but are harder than materials like steel, iron etc. Plenty of raw materials are available for the manufacture. They are durable for number of years.

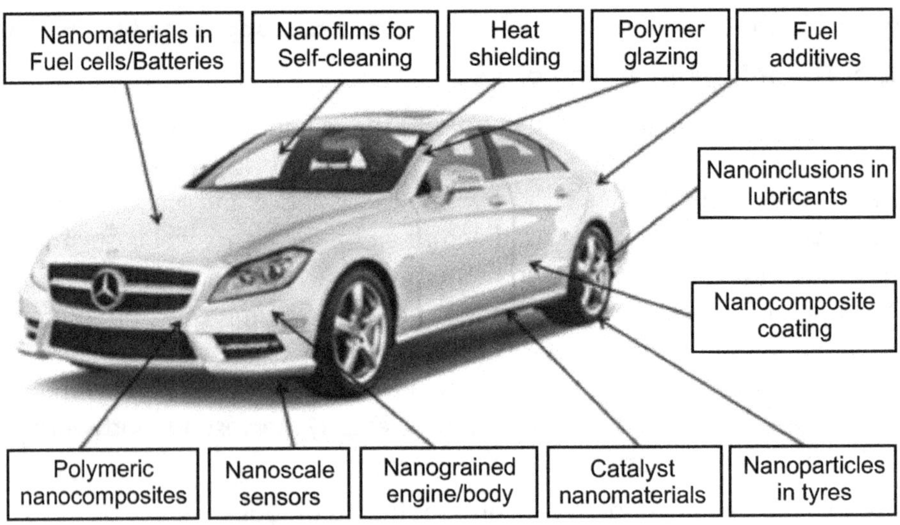

Fig. 6.4: Applications of nanomaterials in cars

6.3.5 Nanotechnology in Space

Following are the applications of nanotechnology in space flight:

Employing materials made from carbon nanotubes to reduce the weight of spaceships like the one shown in Fig. 6.5 (a) while retaining or even increasing the structural strength.

Using carbon nanotubes to make the cable needed for the space elevator, a system which could significantly reduce the cost of sending material into orbit. Nova has a nice video explaining the concepts.

Including layers of bio-nano robots in spacesuits. The outer layer of bio-nano robots would respond to damages to the spacesuit, for example to seal up punctures. An inner layer of bio-nano robots could respond if the astronaut was in trouble, for example by providing drugs in a medical emergency. Deploying a network of nanosensors to search large areas of planets such as Mars for traces of water or other chemicals.

Producing thrusters for spacecraft that use MEMS devices to accelerate nanoparticles. This should reduce the weight and complexity of thruster systems used for interplanetary missions. One cost-saving feature of these type of thrusters is their ability to draw on more or less of the MEMS devices depending upon the size and thrust requirement of the spacecraft, rather than designing and building different engines for different size spacecrafts.

Using carbon nanotubes to build lightweight solar sails that use the pressure of light from the sun reflecting on the mirror-like solar cell to propel a spacecraft. This solves the problem of having to lift enough fuel into orbit to power spacecraft during interplanetary missions.

Working with nanosensors to monitor the levels of trace chemicals in spacecraft to monitor the performance of life support systems.

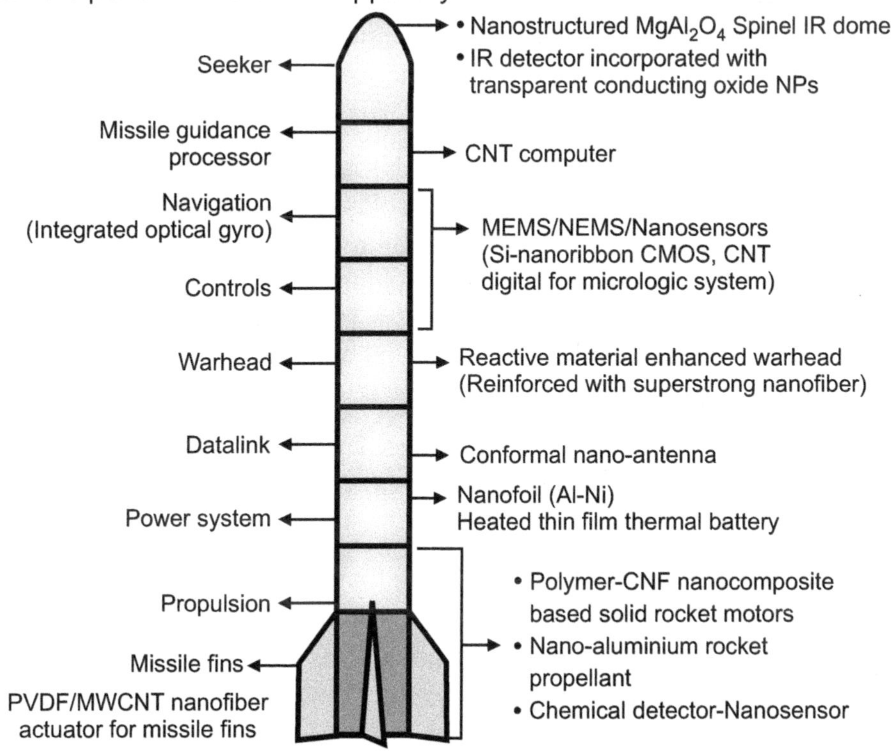

Fig. 6.5 (a): Use of nanomaterials in space jet

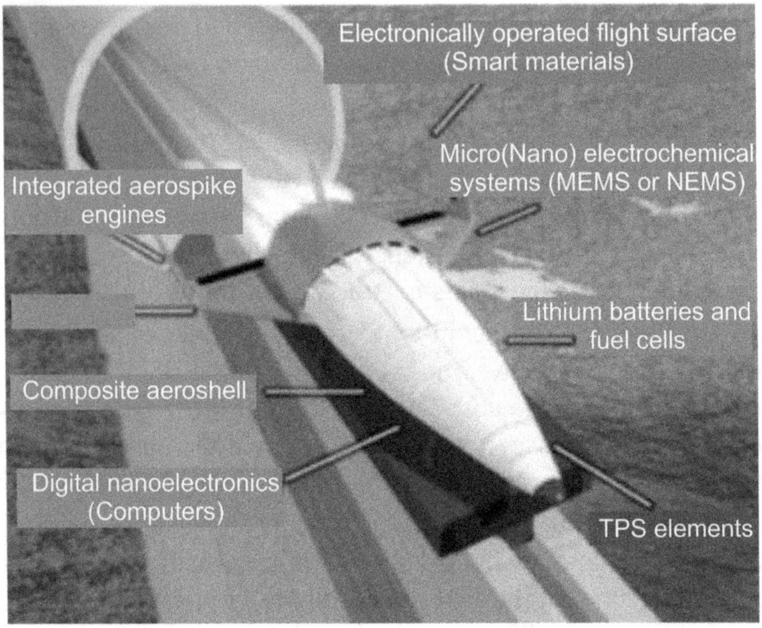

Fig. 6.5 (b): Photo of space jet

6.3.6 Nanotechnology in Defense

Military types haven't failed to notice that nanotechnology could make a big difference to how troops function, travel, and stay safe. The options range from what the well-dressed soldier will wear to methods for making aircraft that can change shape while the plane is in flight.

The Institute for Soldier Nanotechnologies (ISN) is in the business of developing and taking advantage of nanotechnology to help soldiers to survive in battle conditions. A nanobattle suit is being developed that could be as thin as spandex and contain health monitors and communications equipment. Nanomaterials can also provide strength that far surpasses currently available materials, providing bullet shielding that's much more effective. These jumpsuit style outfits might even be able to react to and stop biological and chemical attacks. This protection and these devices would be integrated into one suit that would be more efficient and lightweight than current packs. The U.S. Army Natick Soldier Systems Center has published a white paper that discusses how nanotechnology may be used in the "Future Soldier Initiative".

Researchers have worked on aircraft that swing their wings in close for high-speed flight and extend their wings to provide more lift for takeoff and landing. Unfortunately, the hinges that allow the wings to swing add weight, so researchers are developing materials that will need only an electrical voltage to change the shape of aircraft wings and other structures. NASA has developed a carbon nanotube polymer composite that bends when a voltage is applied.

Soldier Battlesuit: A battlesuit such us that being developed by ISN would be required to remain lightweight and comfortable while stopping bullets, protecting against toxins, monitoring vital signs and administering first aid where possible. Battlesuit research is still in its infancy but has already made some advancement in the fields of communications, strength and soldier protection.

Communications: Just as communications from ships at sea used to utilize coded messages transmitted by means of flashing lights specially coated polymer threads woven into the suit can allow silent communication between soldiers. The system can be tuned to different light wavelengths to prevent eavesdropping or detection by enemy units.

Strength: Polymer molecular muscle ribbons in the suit can magnify a soldier strength by up to ten times. At present the muscles are slow to react and therefore not practical in most battlefield applications.

Protection: Kevlar is already the material of choice for protection against bullets and other ballistics and nanotechnology is being applied to further increase its functionality. Testing is underway on a shock-resistant material five times stronger than steel and more than twice as strong as any other impact-resistant material currently in use. Protection from chemical and biological agents is being provided for with the use of special molecules called dendrimers. The dangerous chemicals stick to dendrimers and are rendered harmless.

Aerospace Applications: In aerospace based defence applications the primary concern is improving strength to weight ratios. As an example, nanotechnology is being applied to aluminium to change phases and microstructure in order to make it perform like titanium – but without the weight.

Coatings: High strength, corrosion resistant coatings are another military use for nanotechnology in order to improve durability, corrosion resistance and reliability. These materials can sense damage or corrosion and automatically initiate repair of some damage. The potential is also there for coatings to change colour when required. This could include adaptive camouflage for tanks moving from jungle to open fields or into urban areas.

Carbon Nanotube Composites: Further to improving strength to weight ratios, several companies are developing high strength, light weight composite materials using carbon nanotubes. Applications for these composites include aircraft wings.

Nanotechnology can help to the military in the following areas:

- Clothing with greater tolerance for temperature changes.
- Increase surveillance for better protection.
- Smaller cameras.
- Cheap, small, and more effective weapons.
- Exploration of the oceans.

- Augmenting human performance
- Scratch resistant surfaces
- Stronger, thinner and cheaper glass
- Change shape of objects, i.e. armor-like fabric
- Coatings that don't degrade (doesn't need repainting)
- Invisibility
- Faster intensive medical help
- Lighter, faster aircraft which use less fuel
- Submarines and planes that can go undetected by radar.

6.3.7 Nanotechnology in Sports

Current nanotechnology applications in the sports arena include:

Increasing the strength of tennis racquets by adding nanotubes to the frames which increases control and power when you hit the ball.

Filling any imperfections in club shaft materials with nanoparticles; this improves the uniformity of the material that makes up the shaft and thereby improving your swing.

Reducing the rate at which air leaks from tennis balls so that the balls keep their bounce longer.

Using nano-sized material to fill any voids in golf shafts.

Coating kayaks with an epoxy containing carbon nanotubes reduces the chance of damaging the skin of the kayak.

Using a nano-composite film to keep the air in tennis balls longer.

Strengthening fishing rods without increasing their weight by adding silica nanoparticles to fill spaces between carbon fibers.

Improving the gliding performance of skis with a wax containing a nano-composite.

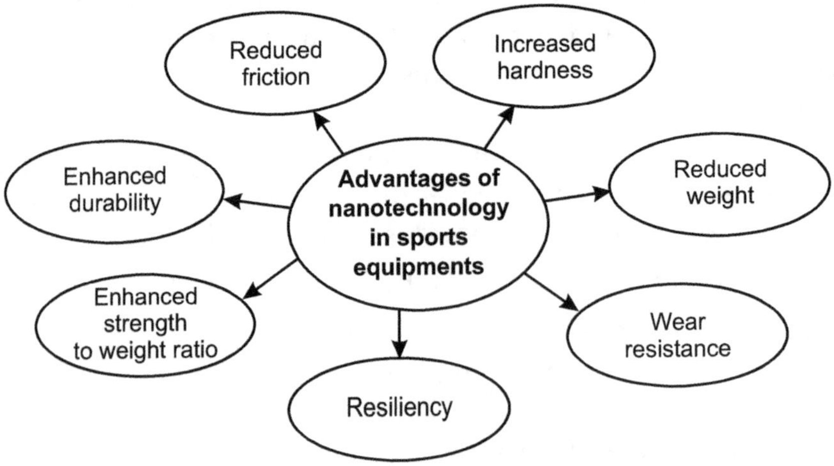

Fig. 6.6: Advantages of nanotechnology in sports equipments

6.3.8 Nanotechnology in Cosmetics

The cosmetics industry is one of those most eager to make the most of the opportunities presented by nanotechnology. Nanomaterials have been used to try and improve the performance of a wide range of products from moisturizer, anti-ageing creams and other skincare products to hair care. There are a number of nano-enhanced products available. The two main uses for nanoparticles in cosmetic products are UV filtering and delivery of active ingredients.

1. **Sunscreens:** Sunscreen that uses titanium dioxide (TiO_2) and zinc oxide (ZnO) block the harmful effects of ultraviolet light while minimizing the white coating on the skin. Skin creams that use proteins derived from stem cells to prevent aging of the skin. These proteins are encapsulated in liposome nanoparticles which merge with the membranes of skin cells to allow delivery of the proteins. Skincare lotions in which nutrients are encapsulated in nanoparticles suspended in liquid, making up a nanoemulsion. This use of nano-sized TiO_2 and ZnO in sunscreens and moisturizers is one of the largest uses of nanomaterials in the cosmetics and personal care markets.
2. **Killing bacteria:** Silver nanoparticles and so-called colloidal silver, which contains silver nanoparticles, are used in many personal care products, because they are very effective at killing bacteria. They are mainly used to ensure cleanliness in equipment, such as make-up instruments, hair brushes, curling tongs, foils for electric razors, foot massagers, tooth brushes, bottle brushes, rubber gloves, hair dryers, hearing aids, facial ionic steamers, and even bidets, though most of these are more likely to be available in Asia. Nano silver is used in soaps, toothpastes, wet wipes, deodorants, lip products, as well as face and body foams.
3. **Smooth moves:** Nanomaterials are used in curling tongs, for example, it is purported to smooth the hair and reduce static through a nanoceramic and silver coating.
4. **Moisturizers and Anti-ageing creams:** 40 years ago, cosmetic companies started using nano-sized parcels of ingredients (also called liposomes or now nanosomes) to improve the solubility of ingredients and add shimmer. Essentially, liposomes are made out of the same material as a cell membrane and act as small capsules or bubbles able to hold and deliver active ingredients and cosmetic materials such as Vitamin E. In the healthcare sector, they are used to deliver therapeutic drugs or vitamins. Nano emulsions are another process in skin creams. Emulsion just means mixing two unblendable liquids (like oil and vinegar in a salad dressing) where one of the liquids is suspended in the other. With nano emulsion, as opposed just plain emulsion, one liquid disperses in nano-scale droplets throughout the other. Nano emulsions, unlike normal emulsions are so fine that they can be sprayed on. Companies that sell them claim that nano emulsions can transport beneficial compounds deep into the skin and in high concentrations. Nano gold is also being used in one moisturizer available for allegedly bringing healing and anti-oxidant properties. Fullerenes, or Bucky balls as they are also known, are carbon molecules that are only about one nanometer in diameter and resemble the structure of some footballs. They are allegedly used for anti-oxidant and smoothing properties in moisturizers.

6.3.9 Nanotechnology in Cloth (Textile) Industry

Nanotechnology has applications in many industries as a new technology. Textile industry is one of the industries that has adopted nanotechnology and benefited from its advantages. The application of nanostructures results in increasing the efficiency of the product and it creates unique properties such as self-cleaning, anti-abrasion, antibacterial, fireproof, UV resistant, anti-stain and anti-shrinkage in the fabrics; therefore, sport and domestic clothes with anti-odour, antibacterial, waterproof, anti-stain, and fireproof properties can be produced by using nanotechnology. Textile industry is very important due to its direct contact with people's lives, and numerous enterprises are active in this field. Textile products can be classified into the following categories:

(a) Clothes (including clothes with specific applications such as laboratory uniforms)
(b) Sport clothes (including shorts, diving suit)
(c) Medical fabrics (including all types of bandages, uniforms)
(d) Domestic and home fabrics
(e) Technical fabrics
(f) Leather and shoes
(g) Nano-coatings

Home fabrics include items such as carpets, bedroom products, curtains, furniture dressings and towels. Fibers, yarns and fabrics with specific applications are classified in technical fabrics group, among which mention can be made of conductive fabrics, non-woven fabrics, UV resistant thread, antibacterial thread and awnings. Nano-coatings include solutions and nanoproducts that are used on the fabrics or clothes either as a final product or as an agent during the completion process.

Two strategies were used to search textile nanoproducts:

1. Search by using nanotechnology in textile industry keywords.
2. Search by using data obtained from technologist enterprises and the related companies.

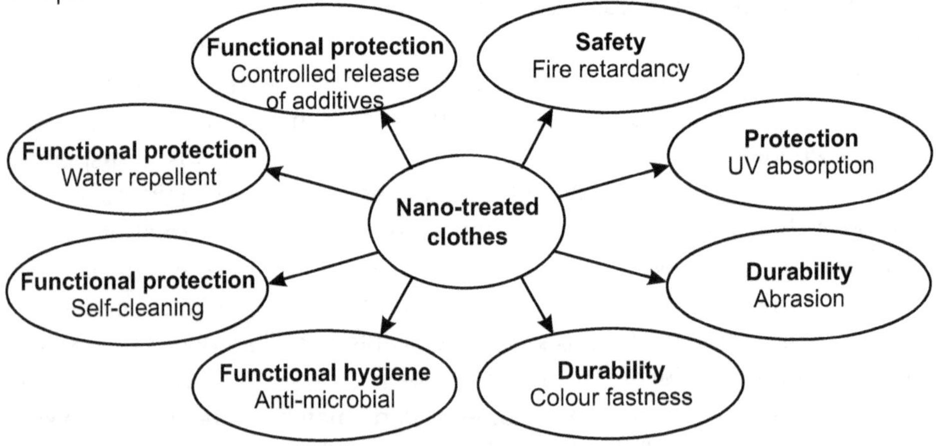

Fig. 6.7: Applications of nanotechnology in cloth industries

EXERCISE

Short-Answer Questions:

1. Applications of nanotechnology depend on what fact/what factors?
2. Enlist the various fields of applications of nanotechnology.
3. Enlist the various fields of applications of nanomaterials.
4. Enlist the various applications of nanomaterials in biology.
5. Enlist the various applications of nanotechnology in automobile industry.
6. Give any two applications of nanotechnology in space flight.
7. For what purpose nanomaterials are used in cosmetics?
8. Give two main uses for nanoparticles in cosmetic products.
9. State various advantages of nanotechnology in sports equipments.
10. In what area nanotechnology can help to the military?
11. What are the classifications of Textile products?

Long-Answer Questions:

1. Write notes on following:
   (a) Applications of nanotechnology,
   (b) Applications of nanomaterials,
   (c) Applications of nanomaterials in electronics and IT,
   (d) Applications of nanomaterials to biology,
   (e) Applications of nanomaterials in medical,
   (f) Applications of nanomaterials in automobile industry,
   (g) Applications of nanomaterials in space,
   (h) Applications of nanomaterials in defense,
   (i) Applications of nanoparticles in cosmetic products,
   (j) Applications of nanomaterials in sports,
   (k) Applications of nanomaterials in Textile (cloth) industries.
2. Describe various applications of nanomaterials in brief.
3. Describe various applications of nanomaterials in electronics and IT.
4. Describe use of nanomaterials in sports equipments in brief.
5. Describe use of nanomaterials in cosmetic products in brief.
6. Describe use of nanomaterials in medical products in brief.
7. Describe various applications of nanomaterials in biology in brief.
8. Explain how nanotechnology is used in defense.
9. What are the applications of nanomaterials in automobile industry?
10. What are the applications of nanomaterials in Textile (cloth) industry?

☆☆☆

PREVIOUS YEARS UNIVERSITY QUESTION PAPERS

OCTOBER 2017

Q. 1. Attempt all of the following:

(a) Define term 'Nanotechnology'.
(b) State the classes of nanomaterials on the basis of structural configurations.
(c) What is bottom-up approach?
(d) Give the names of chemical methods of preparation of nanomaterial.
(e) What is gel?
(f) What is Bragg's Diffraction condition?
(g) Which detectors are used in spectrometers?
(h) Give the Hardness vs Particle size graph for nanomaterial.
(i) Define term 'Nanocrystal'.
(j) State any two applications of nanomaterials in medical field.

Q. 2. Attempt any two of the following:

(a) Write detailed note on the nanomaterials classification on the basis of their origin.
(b) Write a note on UV-visible spectroscopy.
(c) Explain different types of carbon nanotubes.

Q. 3. Attempt any two of the following:

(a) Write a note on 0-D, 1-D, 2-D and 3-D nanomaterials.
(b) State the difference between SEM and TEM.
(c) Describe the structural properties of nanoparticles.

Q. 4. (A) Attempt any one of the following:

(a) Describe the sol-gel method for the preparation of nanoparticles.
(b) (i) What are the advantages and disadvantages of physical vapour deposition process?
(ii) Write a note on the use of nanomaterials in defense.

(B) Attempt any one of the following:

(a) Draw the block diagram of SEM.
(b) State the uses of Nanocrystalline ZnO and TiO_2.

☆☆☆

APRIL 2018

Q. 1. Attempt all of the following:

(a) Define the term 'Nanoscience'.

(b) According to origin, what are the classes of nanomaterials?

(c) What is top-down approach?

(d) State the advantages of top-down processing.

(e) State the biological methods of preparation of nanomaterials.

(f) What is Debye-Scherer's relation?

(g) State any two techniques for the characterization of materials.

(h) Give UV-visible plot for materials with different particle size.

(i) State any two applications of nanomaterials in automobile industry.

(j) What are the different types of carbon nanotubes?

Q. 2. Attempt any two of the following:

(a) Write detailed note on challenges in nanotechnology.

(b) What is electron microscopy? Describe the scanning electron microscope with neat diagram.

(c) Give the applications of Quantum Dots.

Q. 3. Attempt any two of the following:

(a) Write a note on classification of nanomaterials on the basis of structural configuration.

(b) State the different applications of X-ray diffraction method.

(c) Write a note on mechanical properties of nanoparticles.

Q. 4. (A) Attempt any one of the following:

(a) Describe the chemical vapour deposition method for the preparation of nanomaterials.

(b) (i) Explain the terms sol and gel. Describe the advantages of sol and gel process.

(ii) Describe the use of nanomaterials in medical field.

(B) Attempt any one of the following:

(a) Draw the block diagram of Spectrophotometer.

(b) What is Single-walled Nanotube?

☆☆☆

APRIL 2019

Q. 1. Attempt all of the following:

(a) State the names of milestones in the development of nanotechnology.

(b) Define term 'Nanotechnology'.

(c) State hybrid methods of synthesis of nanomaterials.

(d) State any two advantages of the sol-gel process.

(e) Define 'Aerogel'.

(f) State Bragg's law.

(g) State any two different scales to measure hardness.

(h) Define Band Gap Energy.

(i) State any two nanomaterials prominently used in defense field.

(j) State applications of Quantum Dots.

Q. 2. Attempt any two of the following:

(a) Explain mechanism of electrochemical deposition process.

(b) State different types of carbon nanotubes. Explain multiwall nanotubes.

(c) Explain the applications of nanomaterials in field of defense.

Q. 3. Attempt any two of the following:

(a) Write a note on X-ray diffraction technique.

(b) Describe in details of preparation of nanoparticles by colloidal type chemical route method.

(c) What are carbon nanotubes? Explain properties of carbon nanotubes.

Q. 4. (A) Attempt any one of the following:

(a) What is an electron microscopy? Explain working of scanning electron microscope. Give its advantages and disadvantages.

(b) Explain in details :

(i) Challenges in nanotechnology.

(ii) Origin and dimensions of nanomaterials.

(B) Attempt any one of the following:

(a) What is surface and interface effect?

(b) What is sol and gel?

☆☆☆

www.ingramcontent.com/pod-product-compliance
Lightning Source LLC
LaVergne TN
LVHW062210171026
850499LV00044B/1903
* 9 7 8 9 3 8 9 4 0 6 8 0 1 *